Ronald Meester

A Natural Introduction to Probability Theory

Second Edition

Birkhäuser
Basel • Boston • Berlin

Author:

Ronald Meester
Faculteit der Exacte Wetenschappen
Vrije Universiteit
De Boelelaan 1081a
1081 HV Amsterdam
The Netherlands

e-mail: rmeester@cs.vu.nl

2000 Mathematics Subject Classification 60-01

Library of Congress Control Number: 2007942638

Bibliographic information published by Die Deutsche Bibliothek
Die Deutsche Bibliothek lists this publication in the Deutsche Nationalbibliografie;
detailed bibliographic data is available in the Internet at <http://dnb.ddb.de>.

ISBN 978-3-7643-8723-5 Birkhäuser Verlag, Basel – Boston – Berlin

First edition 2003

© 2008 Birkhäuser Verlag AG
Basel · Boston · Berlin
P.O. Box 133, CH-4010 Basel, Switzerland
Part of Springer Science+Business Media
Printed on acid-free paper produced from chlorine-free pulp. TCF ∞

ISBN   978-3-7643-

9 8 7 6 5 4 3 2 1

e-ISBN 978-3-7643-8724-2

www.birkhauser.ch

Contents

Preface to the First Edition   viii

Preface to the Second Edition   x

1   **Experiments**   **1**
    1.1   Definitions and Examples   1
    1.2   Counting and Combinatorics   6
    1.3   Properties of Probability Measures   10
    1.4   Conditional Probabilities   12
    1.5   Independence   18
    1.6   A First Law of Large Numbers   26
    1.7   Exercises   28

2   **Random Variables and Random Vectors**   **35**
    2.1   Random Variables   35
    2.2   Independence   41
    2.3   Expectation and Variance   43
    2.4   Random Vectors   52
    2.5   Conditional Distributions and Expectations   57
    2.6   Generating Functions   62
    2.7   Exercises   64

3   **Random Walk**   **71**
    3.1   Random Walk and Counting   71
    3.2   The Arc-Sine Law   75
    3.3   Exercises   78

4   **Limit Theorems**   **81**
    4.1   The Law of Large Numbers   81
    4.2   The Central Limit Theorem   85
    4.3   Exercises   88

I   Intermezzo   **89**
   I.1   Uncountable Sample Spaces   89
   I.2   An Event Without a Probability?!   90
   I.3   Random Variables on Uncountable Sample Spaces   92

5   Continuous Random Variables and Vectors   **93**
   5.1   Experiments   93
   5.2   Properties of Probability Measures   98
   5.3   Continuous Random Variables   100
   5.4   Expectation   105
   5.5   Random Vectors and Independence   109
   5.6   Functions of Random Variables and Vectors   113
   5.7   Sums of Random Variables   117
   5.8   More About the Expectation; Variance   118
   5.9   Random Variables Which are Neither Discrete Nor Continuous . .   122
   5.10 Conditional Distributions and Expectations   124
   5.11 The Law of Large Numbers   131
   5.12 Exercises   131

6   Infinitely Many Repetitions   **137**
   6.1   Infinitely Many Coin Flips and Random Points in $(0, 1]$ 138
   6.2   A More General Approach to Infinitely Many Repetitions   140
   6.3   The Strong Law of Large Numbers   142
   6.4   Random Walk Revisited   146
   6.5   Branching Processes   147
   6.6   Exercises   151

7   The Poisson Process   **153**
   7.1   Building a Model   153
   7.2   Basic Properties   158
   7.3   The Waiting Time Paradox   162
   7.4   The Strong Law of Large Numbers   164
   7.5   Exercises   165

8   Limit Theorems   **167**
   8.1   Weak Convergence   167
   8.2   Characteristic Functions   169
   8.3   Expansion of the Characteristic Function   173
   8.4   The Law of Large Numbers   176
   8.5   The Central Limit Theorem   179
   8.6   Exercises   180

9   Extending the Probabilities   **183**
   9.1   General Probability Measures   183

A **Interpreting Probabilities**   **187**

B **Further Reading**   **191**

C **Answers to Selected Exercises**   **193**

Index   **195**

Preface to the First Edition

According to Leo Breiman (1968), probability theory has a right and a left hand. The right hand refers to rigorous mathematics, and the left hand refers to 'probabilistic thinking'. The combination of these two aspects makes probability theory one of the most exciting fields in mathematics. One can study probability as a purely mathematical enterprise, but even when you do that, all the concepts that arise do have a meaning on the intuitive level. For instance, we have to define what we mean exactly by independent events as a mathematical concept, but clearly, we all know that when we flip a coin twice, the event that the first gives heads is independent of the event that the second gives tails.

Why have I written this book? I have been teaching probability for more than fifteen years now, and decided to do something with this experience. There are already many introductory texts about probability, and there had better be a good reason to write a new one. I will try to explain my reasons now.

The typical target that I have in mind is a first year student who wants or needs to learn about probability at a stage where he or she has not seen measure theory as yet. The usual and classical way to set things up in this first year, is to introduce probability as a measure on a sigma-algebra, thereby referring to measure theory for the necessary details. This means that the first year student is confronted with a theory of probability that he or she cannot possibly understand at that point. I am convinced that this is not necessary.

I do not (of course) claim that one should not base probability theory on measure theory later in the study, but I do not really see a reason to do this in the first year. One can – as I will show in this book – study discrete and continuous random variables perfectly well, and with mathematical precision, within the realm of Riemann integration.

It is not particularly difficult to write rigorously about discrete probability, but it is harder to do this more generally. There are several texts available which do promise this (no measure theory and rigorous) but I don't think that any of these texts can really say to have kept its promise. I have achieved precision without measure theory, by deviating from the classical route of defining probability measures on sigma-algebras. In this book, probabilities of events are defined as soon as a certain (Riemann) integral exists. This is, as I will show, a very natural thing to do. As a result, everything in this book can be followed with no more background than ordinary calculus.

As a result of my approach, it will become clear where the *limits* of this approach are, and this in fact forms the perfect *motivation* to study measure theory and probability theory based on measure theory later in the study. Indeed, by the end of the book, the student should be dying to learn more about measure theory.

Hence this approach to probability is fully consistent with the way mathematics works: first there is a theory and you try to see how far this gets you, and when you see that certain (advanced) things cannot be treated in this theory, you have to extend the theory. As such, by reading this book, one not only learns a lot about probability, but also about the way mathematics is discovered and developed.

Another reason to write this book is that I think that it is very important that when students start to learn about probability, they come to interesting results as soon as possible, without being disturbed by unnecessary technical complications. Probability theory is one of those fields where you can derive very interesting, important and surprising results with a minimum of technical apparatus. Doing this first is not only interesting in itself. It also makes clear what the limits are of this elementary approach, thereby motivating further developments. For instance, the first four chapters will be concerned with discrete probability, which is rich enough as to say everything about a finite number of coin flips. The very surprising arc-sine law for random walks can already be treated at this point. But it will become clear that in order to describe an *infinite* sequence of such flips, one needs to introduce more complicated models. The so called weak law of large numbers can be perfectly well stated and proved within the realm of discrete probability, but the strong law of large numbers cannot.

Finally, probability theory is one of the most useful fields in mathematics. As such, it is extremely important to point out what exactly we do when we model a particular phenomenon with a mathematical model involving uncertainty. When can we safely do this, and how do we know that the outcomes of this model do actually say something useful? These questions are addressed in the appendix. I think that such a chapter is a necessary part of any text on probability, and it provides a link between the left and right hand of Leo Breiman.

A few words about the contents of this book. The first four chapters deal with *discrete probability*, where the possible outcomes of an experiment form a finite or countable set. The treatment includes an elementary account on random walks, the weak law of large numbers and a primitive version of the central limit theorem. We have also included a number of confusing examples which make clear that it is sometimes dangerous to trust your probabilistic intuition.

After that, in the Intermezzo, we explain why discrete probability is insufficient to deal with a number of basic probabilistic concepts. Discrete probability is unable to deal with infinitely *fine* operations, such as choosing a point on a line segment, and infinitely many *repetitions* of an operation, like infinitely many coin flips.

After the Intermezzo, Chapter 5 deals with these infinitely fine operations, a subject which goes under the name *continuous probability*. In Chapter 6 we continue with infinitely many repetitions, with applications to branching processes, random walk and strong laws of large numbers.

Chapter 7 is devoted to one of the most important stochastic processes, the Poisson process, where we shall investigate a very subtle and beautiful interply between discrete and continuous random variables.

In Chapter 8, we discuss a number of limit theorems based on characteristic functions. A full proof of the central limit theorem is available at that point. Finally, in Chapter 9, we explore the limitations of the current approach. We will outline how we can extend the current theory using measure theory. This final chapter provides the link between this book and probability based on measure theory.

Previous versions of this book were read by a number of people, whose comments were extremely important. I would very much like to thank Hanneke de Boer, Lorna Booth, Karma Dajani, Odo Diekmann, Massimo Franceschetti, Richard Gill, Klaas van Harn, Rob Kooijman, Rein Nobel, Corrie Quant, Rahul Roy, Jeffrey Steif, Freek Suyver, Aad van der Vaart and Dmitri Znamenski for their valuable comments. In particular, I would also like to thank Shimrit Abraham for doing a large number of exercises.

Ronald Meester, Amsterdam, Summer 2003

Preface to the Second Edition

Teaching from a book leads to many remarks and complaints. Despite the care given to the first edition, it turned out that it still contained a fair number of small mistakes. I thank Corrie Quant and Karma Dajani (again), Rob van den Berg, Misja Nuyens and Jan Hoogendijk for their corrections and other remarks that helped me very much in preparing this second edition. Apart from correcting mistakes and typos, I also added a significant number of exercises and some examples. Finally, here and there I slightly rearranged the material. There are, however, no essential differences between the first and second edition.

Ronald Meester, Amsterdam, Fall 2007

Chapter 1

Experiments

We start our investigations with a number of elementary examples which involve probability. These examples lead to a definition of an *experiment*, which consists of a space of possible outcomes, together with an assignment of probabilities to each of these outcomes. We define and study basic concepts associated with experiments, including combinatorics, independence, conditional probabilities and a first law of large numbers.

1.1 Definitions and Examples

When we toss a coin, and have no particular reason to favour heads or tails, then anybody will be happy to assign probability $1/2$ to each of the two possible outcomes. Why is this a reasonable choice? Well, in the first place, if the coin is fair, then heads and tails should receive the *same* probability. Fine, but why do we choose probability $1/2$ each, and not, say $1/10$? The reason for this has to do with our intuitive notion of what a probability means. If we toss the coin k times, and the number of heads among these k tosses is k_h, then the *relative frequency* of heads is equal to k_h/k. Now when k is large, then we tend to think about k_h/k as being close to the probability of heads. The relative frequency of tails can be written as k_t/k, where k_t is the number of tails among the k tosses, and we again think of k_t/k as being close to the probability of tails. Since

$$\frac{k_h}{k} + \frac{k_t}{k} = 1,$$

we see that, at least intuitively, the probabilities of heads and tails should add up to one.

In this simple example, the two possible outcomes form the set

$$\Omega = \{\text{head, tail}\},$$

and we can formalise the assignment of probabilities by defining a function

$$p : \Omega \to [0, 1]$$

given by

$$p(\text{head}) = p(\text{tail}) = 1/2.$$

More generally, if we want to select a number from the set $\Omega = \{1, 2, \ldots, N\}$ without any preference, then we can model this with a function $p : \Omega \to [0, 1]$ defined by

$$p(i) = \frac{1}{N}, \quad i = 1, \ldots, N,$$

that is, we assign probability $1/N$ to each possible outcome. If someone now asks about the probability of seeing either a 1 or a 2, we simply add up the probabilities of 1 and 2 to obtain $2/N$ (assuming of course that $N \geq 2$).

To see that this *adding* of probabilities is reasonable, imagine that we choose a number between 1 and N k times. Denoting the number of 1s and 2s among these k numbers by k_1 and k_2 respectively, then, as discussed above, the relative frequency k_1/k should be close to the probability of a 1. Similarly, the relative frequency of seeing a 1 or a 2, should be close to the probability of seeing a 1 or a 2. Since the relative frequency of a 1 or a 2 is equal to $(k_1 + k_2)/k$, this leads to the conclusion that the probability of seeing a 1 or a 2 should indeed be the sum of the probabilities of 1 and 2.

In these first two examples, all outcomes have the same probability. It is, however, not always appropriate to assign equal probability to all possible outcomes. For instance, suppose that we have an urn which contains three red and two blue balls, so that balls with the same colour are indistinguishable. When we take a ball from the urn without looking, and we are only interested in the colour, then the possible outcomes of this experiment are 'red' and 'blue'. The probability of 'red' should, of course, be 3/5, and the probability of 'blue' should be 2/5.

Another point of concern is that the number of possible outcomes need not be finite. For instance, if you would like to make a probabilistic statement about people's favourite number, then it is not appropriate to take a finite sample space, since there is no a priori bound on this number. In such cases it is often natural to take $\Omega = \mathbb{N} = \{0, 1, 2, \ldots\}$.

In the first four chapters of this book, we shall only be concerned with probabilistic experiments with a finite or *countably infinite* number of possible outcomes. Loosely speaking, an infinite set S is countably infinite if we can list the elements in S as $S = \{s_1, s_2, \ldots\}$. The formal definition is as follows.

Definition 1.1.1. A set S is said to be *countably infinite* if there is a one-to-one correspondence between $\mathbb{N}$ and S. A set which is either finite or countably infinite is said to be *countable*.

For instance, the set $\mathbb{Z}$ of all integers is countable since we can list all elements in $\mathbb{Z}$ as

$$\mathbb{Z} = \{0, -1, 1, -2, 2, \ldots\}.$$

The reason that we stick to countably many outcomes, at least for the time being, is that we do not want to be disturbed by purely technical problems at the beginning of the development. Most (but not all!) probabilistic concepts are already very interesting in this reduced context, and we think that it is better to obtain some probabilistic intuition before going into so called continuous probability, which will have to wait until Chapter 5.

The above examples lead to the following formal definition of an experiment.

Definition 1.1.2. An *experiment* is a countable set of outcomes Ω, called the *sample space*, together with a function $p : \Omega \to [0,1]$, with the property that

$$\sum_{\omega \in \Omega} p(\omega) = 1.$$

The function p is called a *probability mass function*. For $A \subseteq \Omega$, we define the *probability of A* by

$$P(A) = \sum_{\omega \in A} p(\omega),$$

where an empty sum is defined to be 0. A subset of Ω is called an *event*. The map P from the collection of subsets of Ω into $[0,1]$ is called a *probability measure*.

Example 1.1.3 (Throwing a die). When we throw a die, the obvious choice of the sample space is $\Omega = \{1, 2, 3, 4, 5, 6\}$, and the probability mass function should be given by $p(i) = 1/6$, $i = 1, \ldots, 6$. The probability of the event $\{2, 4, 6\}$ that the outcome is even is now easily seen to be

$$P(\{2, 4, 6\}) = p(2) + p(4) + p(6) = \frac{1}{2}. \qquad \square$$

Example 1.1.4. Let us look at families with two children. What is the probability that there is at least one son? To give an answer to this question, we have to model this with an experiment. An appropriate sample space is

$$\Omega = \{BB, BG, GB, GG\},$$

where for instance the outcome BG corresponds to the eldest child being a son, and the youngest a girl. It is reasonable (but see the next exercise) to assign equal probability to all four outcomes. The probability of having at least one son is now the probability of the event $\{BB, BG, GB\}$ and this probability clearly is $3/4$.
$\square$

♠ **Exercise 1.1.5.** What would be your response to someone who takes

$$\Omega = \{BB, BG, GG\},$$

where the outcomes represent two boys, one boy, and no boy respectively, and who assigns equal probability to these *three* outcomes?

Proposition 1.1.6. *It is the case that $P(\Omega) = 1$ and that $P(\emptyset) = 0$.*

♠ **Exercise 1.1.7.** Prove this proposition.

Example 1.1.8. Suppose we want to make probabilistic statements about the number of sons and daughters in families. A possible model is the following. The sample space can be taken as $\Omega = \{(i,j), i,j \in \mathbb{N}\}$, where the outcome (i,j) corresponds to i sons and j daughters. A possible choice for the probability mass function is

$$p((i,j)) = 2^{-i-j-2}.$$

To see that this indeed is a probability mass function, we write

$$\sum_{i=0}^{\infty}\sum_{j=0}^{\infty} 2^{-i-j-2} = \sum_{i=0}^{\infty} 2^{-i-1} \sum_{j=0}^{\infty} 2^{-j-1}$$
$$= 1.$$

What is the probability that there are no sons? The event that there are no sons can be written as $A = \{(0,0),(0,1),(0,2),\ldots\}$ and the probability $P(A)$ of this event is

$$P(A) = \sum_{j=0}^{\infty} p((0,j)) = \sum_{j=0}^{\infty} 2^{-j-2} = 1/2. \qquad \square$$

♠ **Exercise 1.1.9.** Do you think that this is a realistic probability mass function? Why?

The probability of the outcome ω can be written in terms of the probability mass function as $p(\omega)$ or in term of the probability measure as $P(\{\omega\})$. The difference is that p assigns a probability to an *element* of Ω, whereas P assigns a probability to a *subset* of Ω. However, we shall often abuse notation, and write $P(\omega)$ instead of $P(\{\omega\})$.

A useful way to think about an experiment is to imagine that you 'pick a point' from Ω in such a way that the probability to pick ω is just $P(\omega)$. In this interpretation, we say that $P(A)$ is the probability that A *occurs*, that is, $P(A)$ should be thought of as the probability that the chosen ω is a point in A.

Definition 1.1.10. Let A and B be events.

1. We write A^c for the event $\Omega \backslash A$, that is, the event that A does not occur:

$$A^c = \{\omega : \omega \notin A\}.$$

The event A^c is called the *complement* of A.

2. We write $A \cap B$ for the event that A and B both occur, that is,

$$A \cap B = \{\omega : \omega \in A \text{ and } \omega \in B\}.$$

The event $A \cap B$ is called the *intersection* of A and B.

3. We write $A \cup B$ for the event that at least one of the events A or B occurs, that is
$$A \cup B = \{\omega : \omega \in A \text{ or } \omega \in B\}.$$

The event $A \cup B$ is called the *union* of A and B.

We sometimes write $A \backslash B$ for $A \cap B^c$, that is, the event that A occurs, but B does not. When two events A and B have no common element, that is, $A \cap B = \emptyset$, we say that A and B are *disjoint*. A collection of events $A_1, A_2, \ldots$ is called *(pairwise) disjoint*, if $A_i \cap A_j = \emptyset$ for all $i \neq j$.

Example 1.1.11. Consider Example 1.1.3, and let A be the event that the outcome is even, and B be the event that the outcome is at most 4. Then $A = \{2, 4, 6\}$, $B = \{1, 2, 3, 4\}$, $A \cup B = \{1, 2, 3, 4, 6\}$, $A \cap B = \{2, 4\}$ and $A^c = \{1, 3, 5\}$. □

Very often it is the case that all possible outcomes of an experiment have the same probability, see for instance Example 1.1.3 and Example 1.1.4 above. In such cases we can compute the probability of an event A by counting the number of elements in A, and divide this by the number of elements in Ω. This is precisely the content of the following proposition. We denote the number of elements in a set A by $|A|$.

Proposition 1.1.12. *Let Ω be a finite sample space, and let P be a probability measure that assigns equal probability to each outcome. Then*
$$P(A) = \frac{|A|}{|\Omega|},$$

for all $A \subseteq \Omega$.

Proof. Since $P(\omega)$ is the same for all $\omega \in \Omega$, and Ω contains $|\Omega|$ elements, we have that $P(\omega) = |\Omega|^{-1}$, for all $\omega \in \Omega$. Hence,
$$\begin{aligned} P(A) &= \sum_{\omega \in A} P(\omega) \\ &= \sum_{\omega \in A} \frac{1}{|\Omega|} = \frac{|A|}{|\Omega|}. \end{aligned}$$
□

Example 1.1.13. Consider Example 1.1.3 again. The probability of the event $B = \{1, 2, 3, 4\}$ is now easily seen to be 4/6=2/3. Similarly, the probability of $A = \{2, 4, 6\}$ is simply 3/6=1/2. □

Example 1.1.14. Consider Example 1.1.4 again. The probability that a family has a sun and a daughter is the probability of the event $\{BG, GB\}$, and according to Proposition 1.1.12, this probability is equal to 1/2. □

♠ **Exercise 1.1.15.** Show that in a countably *infinite* sample space, it is impossible to assign equal probability to all possible outcomes. (What would happen if this common probability were positive? What would happen if this common probability were zero?) As a consequence, it is impossible to choose a natural number randomly, in such a way that all numbers have the same probability.

The previous examples suggest that it typically is easy to apply Proposition 1.1.12. Indeed, in this example, there were only few outcomes possible, and the counting was very simple. However, counting the number of elements in a set is not always an easy matter, and sometimes relies on tricky arguments. We therefore devote a separate section to methods of counting in various situations, with applications to experiments where all outcomes are equally likely.

1.2   Counting and Combinatorics

In combinatorics, we distinguish between *ordered* and *unordered* sets. In an ordered set, the order plays a role, whereas in an unordered set, it does not. For instance, the list of all ordered subsets of size two of $\{1, 2, 3\}$ consists of $(1, 2), (2, 1), (1, 3), (3, 1), (2, 3)$ and $(3, 2)$; the list of unordered subsets of size two consists of $\{1, 2\}, \{1, 3\}$ and $\{2, 3\}$. The set $\{2, 1\}$ is the same as $\{1, 2\}$ and therefore not listed separately. Note that we write ordered sets between round brackets, and unordered subsets between curly brackets. When we talk about a 'subset', without prefix, then we mean an unordered subset; when the ordering is important this will always be mentioned specifically.

All theory of this section is contained in the following basic result. After the proposition and its proof, we give a number of examples.

Proposition 1.2.1 (Combinatorial counting methods).

(a) *The number of sequences of length k with n symbols is n^k.*

(b) *The number of* ordered *subsets of k elements from a set with n elements is equal to*

$$n \times (n - 1) \times \cdots \times (n - k + 1).$$

In particular, there are $n!$ ways to order a set of n elements.

(c) *The number of subsets of k elements from a set with n elements, denoted by $\binom{n}{k}$, is equal to*

$$\binom{n}{k} = \frac{n \times (n - 1) \times \cdots \times (n - k + 1)}{k!} = \frac{n!}{k!(n - k)!}.$$

This number is called 'n choose k'.

Proof. For (a), note that for the first element of the sequence we have n possibilities. For the second element, we again have n possibilities, and combined with the first element, this gives n^2 possibilities for the first two elements. Continue now in the obvious way.

For (b), note that for the first element we have n possibilities. After choosing the first element, we have only $n - 1$ possibilities for the second, etcetera.

For (c), observe that we ask here for the number of subsets, and then the order is irrelevant. Each collection of k elements can be obtained in $k!$ different orderings, so to obtain the number of subsets of size k, we need to take the number of *ordered* subsets, and divide by $k!$, that is

$$\frac{n \times (n - 1) \times \cdots \times (n - k + 1)}{k!}.$$

$\square$

Example 1.2.2 (Drawing with replacement). Consider an urn with eight balls, numbered $1, \ldots, 8$. We draw three balls *with replacement*, that is, after drawing a ball, we note its number and put it back into the urn, so that it may be drawn a second or even a third time. The sample space Ω of this experiment consists of all sequences of length three, with the symbols $1, \ldots, 8$. According to Proposition 1.2.1(a), Ω has $8^3 = 512$ elements. When we make the assumption that each sequence of length three has the same probability, this leads to the conclusion that any given sequence, for instance $(4, 4, 8)$ has probability $1/512$ to occur. $\square$

Example 1.2.3 (Drawing an ordered collection without replacement). Consider the same urn, with the same balls. We now draw three balls *without replacement*, that is, a chosen ball is *not* put back in the urn. We note the numbers of the chosen balls, in order. The sample space Ω' corresponding to this event is the set consisting of all sequences of length three with the symbols $1, \ldots, 8$, where each symbol can appear at most once. The number of elements in Ω' is the number of *ordered* subsets of size three. According to Proposition 1.2.1(b), this is equal to $8 \times 7 \times 6 = 336$. The probability to see $(3, 7, 1)$ (in this order) is then $1/336$, under the assumption that all outcomes have the same probability. $\square$

Example 1.2.4 (Drawing an unordered subset). Consider the same urn once more, this time choosing three balls simultaneously, so that the order is irrelevant. This experiment corresponds to the sample space Ω'' which consists of all subsets of size three of a set with eight elements. According to Proposition 1.2.1(c), Ω'' has $\binom{8}{3} = 56$ elements. Under the assumption that all outcomes have the same probability, the probability to select the set $\{3, 7, 1\}$ is now $1/56$. Note that this is six times the probability of the event in the previous example. The reason for this is that the set $\{3, 7, 1\}$ can appear in $3! = 6$ different orderings. $\square$

Example 1.2.5 (Master Mind). In the game of Master Mind, there are four gaps in a row, which can be filled with little coloured sticks. There are six different colours for the sticks. One can ask how many combinations there are to do this. This depends on the rules that we agree on.

If we insist that all four gaps are actually filled with different colours, then the number of ways to do this, is the same as the number of ways to choose an ordered set of size four from a set of six elements, that is, $6 \times 5 \times 4 \times 3 = 360$.

If we allow *gaps*, that is, if we allow a gap not to be filled, then the number of possibilities is $7 \times 6 \times 5 \times 4 = 840$.

Finally, if we allow gaps and the same colour to be used more than once, then the number of possibilities is just the number of ordered sequences of length four using seven symbols, that is, $7^4 = 2401$. This last number appears in commercials for Master Mind as the total number of possible combinations. □

♠ **Exercise 1.2.6.** How many possibilities are there when gaps are not allowed, but multiple use of the same colour is?

An application of Proposition 1.2.1 need not always be completely straightforward, and we now give a number of more difficult examples. In the following example, we need to use the proposition twice.

Example 1.2.7. Suppose that we have four math books, five language books and two art books which we want to put on a shelve. If we put them in a completely random order, what is the probability that the books are grouped per subject?

To solve this problem, we first have to decide on the total number of possibilities to put the books on the shelve. An application of Proposition 1.2.1(b) tells us that this number is 11!. How many ways are there to have the books grouped per subject? Well, first of all, there are 3! different orderings of the three subjects. And within the math books we have 4! orderings, and similarly for the other subjects. Hence the total number of orderings so that all books are grouped per subject, is 3!4!5!2!. It follows, using Proposition 1.1.12 that the probability that all books are grouped per subject, is equal to

$$\frac{3!4!5!2!}{11!} = 0.000866.$$

In words this means that if we order the books in some arbitrary way, only a little bit more than 8 out of 10,000 trials will lead to a situation where the books are grouped per subject. □

Example 1.2.8. Suppose that we have four couples who meet for a dinner. They all sit down at a round table, in a completely random manner. What is the probability that no two people of the same sex sit next to each other? To answer this question, note that the number of possible configurations is 8!. Indeed, we can order the 8 chairs around the table in a clockwise manner, say. For the 'first' chair, we have 8 possible people, for the second there are 7 left, etcetera. Among these 8! possible outcomes, we need to count how many have no two people of the same sex next to each other. The first person can be of any sex, but when this person sits down, in the second chair, we have only 4 possibilities left. For the third chair, only 3 people are possible, etcetera. We conclude that the number of configurations with no two people of the same sex next to each other is equal to $8 \times 4 \times 3 \times 3 \times 2 \times 2 \times 1 \times 1$. The probability of the event in question is then, according to Proposition 1.1.12, equal to

$$\frac{8 \times 4 \times 3 \times 3 \times 2 \times 2 \times 1 \times 1}{8!} = \frac{4!3!}{7!}$$

$$= \binom{7}{4}^{-1} = 0.0286.$$

$\square$

Example 1.2.9. When you play bridge , a deck of 52 cards is distributed over four players, denoted by north, south, west and east. Each player receives 13 cards. What is the probability that alls spades end up at north and south? There are several ways to find the answer.

Maybe the most natural thing to do is to look at the exact positions of all 52 cards. We think of each of the four players as having 13 (ordered) positions available and we put one card in each position. Since the order is important, there are 52! different ways to distribute the cards. How many of these distributions lead to a situation in which all spades are either at north or at south? Well, think of 52 possible positions for the cards, 13 at each player, and imagine we first find positions for all the spades. For the first spade, there are 26 positions available, for the second only 25, etcetera. Then, when all the spades have found a position, we can distribute the remaining 39 cards as we like, and this can be done in 39! ways. Hence the total number of distributions with all spades at either south or north is given by $26 \cdot 25 \cdots 15 \cdot 14 \cdot 39!$, and therefore the probability that this happens is

$$\frac{26 \cdot 25 \cdots 15 \cdot 14 \cdot 39!}{52!}.$$

A simple computation shows that this is equal to

$$\frac{26!39!}{13!52!} = \frac{\binom{26}{13}}{\binom{52}{13}}.$$

When you stare at this last formula for a second, you might see a more direct way to arrive at this answer. Indeed, rather than looking at the exact positions of all 52 cards, we could look only at the set of positions taken by the 13 spades. There are $\binom{52}{13}$ of such sets. On the other hand, the number of such sets at either north or south is $\binom{26}{13}$, and this yields the same answer in a much more elegant way. $\square$

The examples show that there is not much theory involved in combinatorics. For each new situation, we have to think again how the counting should be done. Often there are various approaches leading to the correct answer. The only way to get acquainted to combinatorics is to train yourself by doing exercises.

Example 1.2.10 (Fair coin tossing). Suppose we flip a coin n times, where we do not expect the coin to favour head or tail: we say that the coin is *unbiased*. For notational convenience, we shall say that the possible outcomes are 0 and 1, rather than heads and tails. The sample space Ω now consists of all sequences of 0s and

1s of length n. An element of Ω is denoted by $\omega = (\omega_1, \omega_2, \ldots, \omega_n)$, where each ω_i is either 0 or 1.

The total number of such sequences is 2^n (Proposition 1.2.1(a)), and since we assume that the coin is unbiased (that is, does not favour heads or tails), all possible outcomes should be equally likely, and we therefore define a probability mass function by $p(\omega) = 2^{-n}$, for all $\omega \in \Omega$. This means that we can apply Proposition 1.1.12, so to compute the probabilitiy of an event, we need to count the number of elements in this event. This counting method for computing probabilities is not always the best thing to do. (In Section 1.4 we shall encounter better and faster ways for computing certain probabilities.) Here are some examples of the counting method.

Let A be the set

$$A = \{\omega \in \Omega : \omega_1 = 1\},$$

that is, A is the event that the first coin flip yields a tail. The number of elements of A is 2^{n-1}, since the first digit is 1, and the other $n - 1$ digits are unrestricted. Therefore, $P(A) = 2^{n-1}/2^n = 1/2$, as both intuition and common sense require.

A more complicated event is the event B_k, defined as the set of outcomes in which we see k tails and $n - k$ heads. Obviously, we assume that $0 \leq k \leq n$ here. How do we count the number of elements in the set B_k? We need to know how many ways there are to choose k positions for the 1s in a sequence of length n. But this is the same as choosing a subset of size k from a set of size n, and according to Proposition 1.2.1(c), this is simply $\binom{n}{k}$. Hence,

$$P(B_k) = \binom{n}{k} 2^{-n}. \qquad \qquad \square$$

1.3   Properties of Probability Measures

In this section we collect and prove a number of useful properties of probability measures. Throughout the section, the sample space is denoted by Ω and $A, B, \ldots$ are events in Ω.

Lemma 1.3.1.   (a) *For events $A_1, A_2, \ldots$ which are pairwise disjoint, we have*

$$P\left(\bigcup_{i=1}^{\infty} A_i\right) = \sum_{i=1}^{\infty} P(A_i).$$

(b) $P(A^c) = 1 - P(A)$.

(c) *If $A \subseteq B$, then $P(A) \leq P(B)$.*

More precisely, we have that $P(B) = P(A) + P(B \backslash A)$.

(d) $P(A \cup B) = P(A) + P(B) - P(A \cap B)$.

Proof. (a) We have

$$P\left(\bigcup_{i=1}^{\infty} A_i\right) = \sum_{\omega \in \cup_i A_i} P(\omega) = \sum_{\omega \in A_1} P(\omega) + \sum_{\omega \in A_2} P(\omega) + \cdots$$

$$= \sum_{i=1}^{\infty} P(A_i).$$

(b) Take $A_1 = A$, $A_2 = A^c$ and $A_j = \emptyset$, for all $j \geq 3$. It follows from (a) that $1 = P(\Omega) = P(A \cup A^c) = P(A) + P(A^c)$, proving (b).

(c) We can write $B = A \cup (B \backslash A)$. This is a union of disjoint events, and the result now follows from (a).

(d) We can write $A \cup B = A \cup (B \backslash A)$, which is a disjoint union. Hence we find that

$$\begin{aligned} P(A \cup B) &= P(A) + P(B \backslash A) = P(A) + P(B \backslash (A \cap B)) \\ &= P(A) + P(B) - P(A \cap B), \end{aligned}$$

where the last equality follows from (c). □

The property proved in (a) is called *countable additivity* of the probability measure P. It expresses the very intuitive idea that we can add up probabilities of disjoint events, something we already anticipated in Section 1.1. The property in (b) is also very natural: for any event A, either A or its complement A^c occurs, but not both, and therefore their probabilities should add up to 1, as they do. The property in (c) simply states that when you make an event larger, its probability increases. Finally, (d) can be understood intuitively as follows: if you want to add all the probabilities of elements in the union of A and B, you can first add up everything in A, and then add up everything in B. However, by doing that, the elements that are in *both* A and B are counted twice, and we need therefore to subtract the probabilities in the intersection $A \cap B$.

The property in (d) can be generalised to more than two events, as follows.

$$\begin{aligned} P(A \cup B \cup C) &= P((A \cup B) \cup C) \\ &= P(A \cup B) + P(C) - P((A \cup B) \cap C) \\ &= P(A) + P(B) - P(A \cap B) + P(C) - P((A \cap C) \cup (B \cap C)) \\ &= P(A) + P(B) - P(A \cap B) + P(C) \\ &\quad - P(A \cap C) - P(B \cap C) + P(A \cap B \cap C). \end{aligned}$$

Lemma 1.3.2. *For events $A_1, \ldots, A_n$, it is the case that*

$$P\left(\bigcup_{i=1}^{n} A_i\right) = \sum_i P(A_i) - \sum_{i<j} P(A_i \cap A_j) + \sum_{i<j<k} P(A_i \cap A_j \cap A_k) - \cdots$$

$$+ (-1)^{n+1} P(A_1 \cap A_2 \cap \cdots \cap A_n).$$

♠ **Exercise 1.3.3.** Prove this lemma by induction.

♠ **Exercise 1.3.4.** Show that for pairwise disjoint events $A_1, \ldots, A_n$, we have

$$P\left(\bigcup_{i=1}^n A_i\right) = \sum_{i=1}^n P(A_i).$$

This is called *finite additivity* of the probability measure.

1.4 Conditional Probabilities

When we talk and think about probability, the concept of independence plays a crucial role. For instance, when we flip a coin twice, we are inclined to say that the outcome of the first flip 'says nothing' about the outcome of the second. Somehow, we believe that information about the first flip gives us no information about the outcome of the second. We believe that the two outcomes are *independent* of each other.

On the other hand, when we throw a die, and consider the event E_3 that the outcome is equal to 3, and the event $E_{\leq 4}$ that the outcome is at most 4, then information about $E_{\leq 4}$ does, in fact, change the probability of E_3. Indeed, if I tell you that $E_{\leq 4}$ does *not* occur, then we know for sure that E_3 cannot occur either, and hence the new probability of E_3 had better be 0. If I tell you that $E_{\leq 4}$ does occur, then there are four possibilities left. The new probability that E_3 occurs should therefore be $\frac{1}{4}$, see Example 1.4.2 below.

The last argument can be carried out in much more general terms, as follows. Suppose I tell you that in a certain sample space Ω, we have two events A and B, with probabilities $P(A)$ and $P(B)$ respectively. This means that a fraction $P(A)$ of all probability mass is concentrated in the event A, and similarly for B. Now suppose that I know that the event B occurs. Does this new information change the probability of the event A? Well, we now know that only outcomes in B matter, and we can disregard the rest of the sample space. Hence we only need to look at the probabilities of elements in B. The new probability that A occurs should now be the fraction of probability mass in B that is also in A. That is, it should be the sum of the probabilities of all outcomes in $B \cap A$, divided by the probability of B.

There is an alternative way to arrive at the same conclusion. As observed before, we like to interpret probabilities as relative frequencies. Suppose that we repeat a certain experiment k times (where k is large), and on each occasion we observe whether or not the events A and B occur. The number of occurrences of an event E is denoted by k_E. Conditioning on B means that we only look at those outcomes for which B occurs, and disregard all other outcomes. In this smaller collection of trails, the fraction of the outcomes for which A occurs is $k_{A \cap B}/k_B$ which is equal to

$$\frac{k_{A \cap B}/k}{k_B/k},$$

and this should be close to
$$\frac{P(A \cap B)}{P(B)}.$$

Definition 1.4.1. Suppose A and B are events in a sample space Ω, and suppose that $P(B) > 0$. The *conditional probability* of A given B is written as $P(A|B)$ and defined as
$$P(A|B) = \frac{P(A \cap B)}{P(B)}.$$

Example 1.4.2. Suppose we throw a die. What is the conditional probability of seeing a 3, conditioned on the event that the outcome is at most 4? Well, denoting the event of seeing a 3 by E_3, and the event that the outcome is at most 4 by $E_{\leq 4}$, we have, in the obvious sample space, that $P(E_3) = 1/6$, $P(E_{\leq 4}) = 2/3$ and $P(E_3 \cap E_{\leq 4}) = P(E_3) = 1/6$. Hence
$$P(E_3|E_{\leq 4}) = \frac{P(E_3 \cap E_{\leq 4})}{P(E_{\leq 4})} = \frac{1/6}{2/3} = \frac{1}{4},$$
which makes sense intuitively (why?). □

Example 1.4.3. Consider fair coin tossing, and suppose we flip the coin twice. Let A be the event that the *first* coin flip is a 0, and let B be the event that *at least one* of the coin flips gives a 0. Then $A = \{(0,0), (0,1)\}$ and $B = \{(0,0), (0,1), (1,0)\}$. We can compute $P(A|B)$ as follows:
$$\begin{aligned} P(A|B) &= \frac{P(A \cap B)}{P(B)} \\ &= \frac{P((0,0), (0,1))}{P((0,0), (0,1), (1,0))} = \frac{1/2}{3/4} = \frac{2}{3}. \end{aligned}$$
□

Example 1.4.4. Let Ω consist of the twelve months in a year, and denote by L the set of months with 31 days. Furthermore, we denote by R the set of months whose name contains the letter r. We assign probability $1/12$ to each month. Then $P(L) = 7/12$, since there are 7 months with 31 days (check!). Similarly, one can check that $P(R) = 8/12$. Since the intersection of L and R contains only 4 months (namely january, march, october and december) we have that
$$P(R|L) = \frac{P(R \cap L)}{P(L)} = \frac{4}{7}.$$
□

♠ **Exercise 1.4.5.** Compute $P(L|R)$ in the last example.

Conditional probabilities can be quite counterintuitive, as suggested by the following two examples. The first example also contains an important lesson about the choice of the sample space.

Example 1.4.6. Suppose that we investigate families with two children. We assume that boys and girls are equally likely, and that the sex of the children are independent. The experiment of picking a random family with two children now corresponds to the sample space $\Omega = \{BB, BG, GB, GG\}$, where for instance BG indicates the outcome that the first child was a boy and the second a girl. The natural probability measure assigns equal probability to each outcome. The probability that a family has at least one boy is now simply $P(BB, BG, GB) = 3/4$. Now suppose that we obtain information that a family we have chosen has at least one boy. What is the conditional probability that the other child is also a boy? Well, we are asked about $P(A|C)$ where A is the event $\{BB\}$, and C is the event $\{BB, BG, GB\}$. According to the formulas, we find that $P(A|C) = P(A \cap C)/P(C)$ which is simply $P(BB)/(3/4) = 1/3$. This is already slightly counterintuitive perhaps, because one might think that the fact that we know that there is at least one boy, says nothing about the sex of the other child. Yet, it does.

Now suppose that we visit a family with two children, and suppose that we know already that this family has at least one boy. We ring at the door, and a boy opens the door. This, apparently, is no new information. It seems that the only thing that happens now is that our previous knowledge about the existence of at least one boy is confirmed. But this is not true! We have performed a new experiment, and our original sample space $\Omega = \{BB, BG, GB, GG\}$ does no longer suffice to describe this experiment. Indeed, we cannot learn from Ω which child opened the door. What we need to do is enlarge our sample space, perhaps as

$$\Omega' = \{B^*B, BB^*, B^*G, BG^*, G^*B, GB^*, G^*G, GG^*\},$$

where a $*$ refers to the child that opened the door. So for instance BG^* is the event that the first child is a boy, the second a girl, and that the girl opened the door. In this new experiment, it is reasonable to assign equal probability to all possible outcomes again. The new conditional probability that the other child is also a boy can be computed as follows. We want to know the conditional probability of $\{B^*B, BB^*\}$ given $\{B^*B, BB^*, B^*G, GB^*\}$ and this is now simply computed as being equal to $1/2$. Hence, the bare fact that a boy opened the door does change the conditional probability that the other child is also a boy. $\square$

It is important that you realise what happened in this example. In order to give an answer to the second problem, we had to change the sample space. The first sample space Ω was not big enough to contain the various events of the second experiment.

Example 1.4.7. Consider the following situation. There are three people, A, B and C, in a room, and each of these three people gets a hat on their head which is either red or blue, in such a way that all eight possible ways to do this have the same probability $1/8$. Everybody can see the hat of the other two people, but they can not see their own hat. The people in the room are not allowed to talk to each other.

We view A, B and C as a team and we ask the team to select at least one of them who has to guess the colour of his own hat. Before the team entered the room, and before they receive their hats, they can have agreed on a certain strategy. The question is whether or not the team can come up with a strategy which yields only correct guesses with probability *larger* than 1/2. So if the team makes one guess, this guess should be correct; if the team (for whatever reason) decides to select two or three persons to guess their colour, all guesses must be correct for the team to win.

Note that a strategy with succes probability 1/2 is easy to find: the team can decide that A will always be the one who will make a guess, and then A simply simply flips a coin to decide between red and blue. The probability that A guesses correctly is then of course 1/2. In fact, it is hard to imagine that one can improve on this, since whatever strategy is used, someone has to make a guess, and once it has been decided that B, say, makes a guess, B, not knowing his own colour, will make the wrong guess with probability 1/2.

Convincing as this may sound, it is not true. In fact, there is a strategy that yields the correct colour with probability as big as 3/4. Let me first describe the strategy, and then explain why the above reasoning is wrong.

The team can decide on the following strategy: each of the three people A, B and C do the following. They look at the hats of the other two people. If these two hats have the same colour, then he (or she) does make a guess and guesses the *other* colour. Why does this work? Suppose that not all hats have the same colour, for instance A is red, B is red, and C is blue. In such a case only C makes a guess, and his guess will be correct. So unless all hats have the same colour, this strategy gives the right answer. If all hats are red, say, then all three people make a guess, and they will all be wrong, and hence the team loses. But when the colours are any of RRB, RBR, RBB, BRB, BBR or BRR (in the order of A, B and C), the team wins with this strategy. Hence the probability to win is 6/8=3/4.

All right, so we have this strategy with succes probability 3/4. But it is still not so clear what was wrong with the original reasoning which seemed to tell us that one can never get a succes probability bigger than 1/2. The key to understand this paradox has everything to do with conditional probabilities. In the reasoning above, we first *assumed* that A makes a guess, and given this fact, we claimed that the succes probability cannot be bigger than 1/2. Well, this is correct, even for the present strategy, since the probability that A guesses correctly, *given that A makes a guess*, is equal to

$$\frac{P(\text{A makes a guess and guesses correctly})}{P(\text{A makes a guess})} = \frac{P(BRR, RBB)}{P(BRR, RBB, BBB, RRR)}$$

$$= \frac{1}{2}.$$ □

Conditional probabilities are very useful for computing unconditional probabilities of events, by using the following result.

Definition 1.4.8. We call a countable collection of events $B_1, B_2, \ldots$ a *partition* of Ω if $B_1, B_2, \ldots$ are pairwise disjoint and satisfy

$$\bigcup_i B_i = \Omega.$$

Example 1.4.9. Let $\Omega = \{1, 2, \ldots, 6\}$. Then $B_1 = \{1, 3, 5\}$ and $B_2 = \{2, 4, 6\}$ form a partition of Ω.   □

Example 1.4.10. Let Ω be the collection of all sequences of length n consisting of 0s and 1s (see Example 1.2.10). Then $A = \{\omega = (\omega_1, \ldots, \omega_n) \in \Omega : \omega_1 = 1\}$ and $B = \{\omega = (\omega_1, \ldots, \omega_n) \in \Omega : \omega_1 = 0\}$ form a partition of Ω.   □

Theorem 1.4.11. *Let $B_1, B_2, \ldots$ be a partition of Ω such that $P(B_i) > 0$ for all i, and let A be any event. Then*

$$P(A) = \sum_i P(A|B_i)P(B_i).$$

Proof. We can write A as a disjoint union,

$$A = (A \cap B_1) \cup (A \cap B_2) \cup \cdots.$$

It then follows from Lemma 1.3.1 that

$$P(A) = \sum_i P(A \cap B_i) = \sum_i P(A|B_i)P(B_i).   \qquad □$$

Here is an example from which the power of the previous theorem becomes clear.

Example 1.4.12. Suppose we have a population of people. Suppose in addition that the probability that an individual has a certain disease is $1/100$. There is a test for this disease, and this test is 90% accurate, in the sense that the probability that a sick person is tested positive is 0.9, and that a healthy person is tested positive with probability 0.1. One particular individual is tested positive. Perhaps this individual is inclined to think the following: 'I have been tested positive by a test which is accurate 90% of the time, so the probability that I have the disease is 0.9.' However, this is not correct. Indeed, we ask for the conditional probability to have the disease, given that the test is positive, while the assumptions deal with the conditional probability of a positive test, given the disease. In general, it is certainly not true that $P(A|B)$ is the same as $P(B|A)$.

Now, let A be the event that this individual has the disease, and let B be the event that the test is positive. The individual is interested in the conditional probability $P(A|B)$. The assumptions tell us that given that a person is sick, the test is positive with probability 0.9. Hence the assumptions tell us that $P(B|A) = 0.9, P(B|A^c) = 0.1$ and that $P(A) = 0.01$. To compute $P(A|B)$ we proceed as follows, applying the previous theorem in the computation of $P(B)$ in the denominator.

$$P(A|B) = \frac{P(A \cap B)}{P(B)} = \frac{P(B|A)P(A)}{P(B|A)P(A) + P(B|A^c)P(A^c)}$$

$$= \frac{0.9 \cdot 0.01}{0.9 \cdot 0.01 + 0.1 \cdot 0.99} = 0.09.$$

Hence the probability that a person who is tested positively actually has the disease is equal to 0.09. This seems quite absurd, but the underlying reason is not hard to grasp. Since the disease is rare, there are many more people without the disease than with the disease. Therefore, it is much more likely that an individual is healthy and is tested positive, than that he or she is sick and the test is correct. $\square$

The computation in the last example is an example of the following general result.

Theorem 1.4.13 (Bayes' rule). *Let $B_1, B_2, \ldots, B_n$ be a partition of Ω such that $P(B_i) > 0$ for all i, and let A be any event with $P(A) > 0$. Then for all i,*

$$P(B_i|A) = \frac{P(A|B_i)P(B_i)}{\sum_{j=1}^{n} P(A|B_j)P(B_j)}.$$

Proof. We write

$$P(B_i|A) = \frac{P(A \cap B_i)}{P(A)}$$

$$= \frac{P(A|B_i)P(B_i)}{\sum_{j=1}^{n} P(A|B_j)P(B_j)},$$

according to the definition and Theorem 1.4.11. $\square$

This rule can be used in a situation where we know the probabilities of the B_i's, and then learn something about the occurrence of an event A. Does this new information change the probabilities of the B_i's? In other words, is $P(B_i|A)$ different from $P(B_i)$? Bayes' rule gives a way to compute $P(B_i|A)$. See the forthcoming Example 1.5.16 for another interesting example of applying Bayes' rule.

Example 1.4.14. Suppose that we have two urns, urn I containing two white balls and three blue balls, and urn II containing three blue balls. We pick a random ball from urn I, put it into urn II and after that pick a random ball from urn II. What is the probability that the second ball is blue? To answer this question, we can define the corresponding sample space Ω as

$$\Omega = \{bb, bw, wb, ww\},$$

where bb refers to the outcome of two blue balls, etcetera. How do we assign probabilities to the outcomes in Ω? At this point, it turns out that in order to assign probabilities to events, we in fact will need Theorem 1.4.11. Let A be the event that the final ball is blue, that is, $A = \{bb, wb\}$, and let B be the event that the first ball is blue, that is, $B = \{bb, bw\}$. Clearly, we want $P(B) = 3/5$. Also, when the first ball is blue, this blue ball is put into urn II which after that contains four blue balls. This means that the second ball must necessarily be blue, and we find that it should be the case that

$$P(A|B) = 1.$$

Similarly, we should have

$$P(A|B^c) = \frac{3}{4},$$

since if the first ball is white, then urn II contains one white and three blue balls. Hence, Theorem 1.4.11 tells us that we have

$$P(A) = P(A|B)P(B) + P(A|B^c)P(B^c) = 1 \cdot \frac{3}{5} + \frac{3}{4} \cdot \frac{2}{5} = \frac{9}{10},$$

which is the answer to the original question. We can now also assign probabilities to all outcomes. Indeed, we have

$$P(bb) = P(A \cap B) = P(A|B)P(B) = \frac{3}{5},$$

and similarly for the other outcomes.   □

♠ **Exercise 1.4.15.** Complete the assignment of probabilities in the last example by calculating $P(bw)$, $P(wb)$ and $P(ww)$. Check that they sum up to 1.

The last example contains an important lesson. In this example, we want something to be true for certain conditional probabilities, in this case for instance the probability that the second ball is blue, given that the first ball was blue. We were able to compute probabilities of events *without* first computing the full probability measure on Ω, simply by using the rules of Theorem 1.4.11. This is a very common situation. In many probabilistic problems, there is no need to worry about the sample space or the probability measure, and we can compute probabilities as in the last example. If you want, you can always, after that, compute the full probability measure, but we will not always do that in this book.

1.5   Independence

Now that we have defined the notion of conditional probability, we can define what we mean by independent events. Clearly, for two events A and B to be independent, the probability of A should not change when we learn that B occurs.

Therefore, we could define A and B to be independent whenever $P(A) = P(A|B)$. However, this definition would not be completely satisfactory, for two reasons. First, $P(A|B)$ would not be defined when $P(B) = 0$, and second, the definition would not be symmetric in A and B. Hence we define independence as follows.

Definition 1.5.1. Two events A and B are said to be *(mutually) independent* whenever

$$P(A \cap B) = P(A)P(B).$$

More generally, the events $A_1, A_2, \ldots, A_n$ are called *independent* if

$$P\left(\bigcap_{j \in J} A_j\right) = \prod_{j \in J} P(A_j),$$

for all index sets $J \subseteq \{1, 2, \ldots, n\}$.

Note that if $P(B) > 0$, A and B are independent if and only if $P(A|B) = P(A)$.

Example 1.5.2. We choose a random card from a deck of 52 cards. Let A be the event that the card is a queen, and B be the event that it is a spade. Then $P(A) = 4/52 = 1/13$ and $P(B) = 13/52 = 1/4$. The event $A \cap B$ is the event that we draw a spade queen, the probability of which is just $1/52$. We see that $P(A)P(B) = P(A \cap B)$ and hence the events A and B are independent. In words, this means that if I tell you that I have drawn a queen, you learn nothing about the card being spade or not. □

Example 1.5.3. The events L and R in Example 1.4.4 are not independent, since $P(L) = 7/12$, $P(R) = 8/12$ and $P(L \cap R) = 4/12$. The reason is that overall, 2 out of 3 months have an r in their name, but among the 7 long months, 4 have an r in their name, which means that they are somewhat 'over-represented'. □

♠ **Exercise 1.5.4.** Suppose that $P(B) = 0$ or 1. Show that B is independent of A, for any event A.

♠ **Exercise 1.5.5.** Suppose that A and B are independent. Show that A^c and B are also independent. Show that A^c and B^c are independent. Show that A and B^c are independent.

The following example explains why in the definition of independence for more than two events, we need to require $P(\bigcap_{j \in J} A_j) = \prod_{j \in J} P(A_j)$ for all index sets J, and not only for such sets of size 2.

Example 1.5.6. Consider the sample space

$$\Omega = \{123, 132, 111, 213, 231, 222, 312, 321, 333\},$$

and define a probability measure by assigning probability $1/9$ to each outcome. Define the event A_k as the event that the kth digit is a 1, for $k = 1, 2$ or 3. It is easy to check that $P(A_k) = 1/3$, and $P(A_1 \cap A_2) = P(A_3 \cap A_2) = P(A_1 \cap A_3) = 1/9$. Hence, A_1 and A_2 are independent, and the same is true for the other pairs. However, $P(A_1 \cap A_2 \cap A_3) = 1/9$ and this is not equal to $P(A_1)P(A_2)P(A_3)$. This means that the collection of events A_1, A_2, A_3 is not independent. Since every pair A_i, A_j is independent, we call the family of events *pairwise independent*. This example shows that a family can be pairwise independent without being independent.   □

Example 1.5.7. This example is classical and bound to surprise you. Suppose we have r people in a room. We assume that their birthdays are equally likely to be any day of the year (which we assume to have 365 days, ignoring leap years). Furthermore, we assume that the events that the birthday of a particular person is on a particular day, are independent. What is the probability that no two persons in the room have the same birthday?

    We can order the persons from 1 to r, and convince ourselves, using Proposition 1.2.1, that there are $(365)^r$ possible collections of birthdays. Now let the event that the birthday of person i is day y be denoted by $E(i, y)$, for $i = 1, \ldots, r$ and $y = 1, \ldots, 365$. By independence, the probability of any outcome $\cap_{i=1}^r E(i, y_i)$, is equal to the product of the individual probabilities:

$$P\left(\bigcap_{i=1}^r E(i, y_i)\right) = \prod_{i=1}^r P\left(E(i, y_i)\right) = \left(\frac{1}{365}\right)^r.$$

This means that all possible outcomes of this experiment have the same probability and Proposition 1.1.12 applies. Hence we only need to count the number of outcomes in which no two birthdays coincide. How many outcomes have no common birtday? Well, there is no restriction on the first, but when we know the birthday of person 1, we have only 364 possibilities for the scond, etcetera. Hence we conclude that the probability of having no common birthday is equal to

$$\frac{365 \cdot 364 \cdots (365 - r + 1)}{(365)^r}.$$

Now you can check that this probability is less than $1/2$ for $r = 23$. The very surprising conclusion is that in a collection of 23 people, the probability that at least two of them have the same birthday is larger than $1/2$.   □

♠ **Exercise 1.5.8.** A number of years ago, a mathematician explained the above example on television in a talk show. The host of that particular show did not believe him. He looked around at the audience, about 500 people, and asked: 'My birthday is on the 23rd of November. Is there anyone else who has the same birthday?' No one answered. The mathematician replied that this was very well possible, since he had only made a probabilistic statement. This answer, although not completely wrong, was very misleading. Do you see the correct answer to the point made by the host of the show?

♠ **Exercise 1.5.9.** Before reading the next example, it is a good idea to try to prove the fact that for all real numbers a and b, and all $n \in \mathbb{N}$, we have

$$(a+b)^n = \sum_{k=0}^{n} \binom{n}{k} a^k b^{n-k}.$$

To prove this, you can use the combinatorial interpretation of $\binom{n}{k}$: when you write the product of the n terms on the left, decide in how many ways you can get a factor a^k.

Example 1.5.10 (General coin tossing). We generalise Example 1.2.10 by making the probability of a 1 arbitrary, say p. This implies that it is no longer the case that all possible outcomes have the same probability. Indeed, each coin flip results in a 1 with probability p, and in a 0 with probability $1 - p$. Clearly, we want the outcomes of different flips to be independent. Hence the probability that the first two flips both result in 1, should have probability p^2. This reasoning leads to the conclusion that any outcome with k 1s and $n - k$ 0s, should have probability $p^k(1-p)^{n-k}$. Does this make sense? I mean, is P thus defined indeed a probability measure? To check this, we compute

$$\sum_{\omega \in \Omega} P(\omega) = \sum_{k=0}^{n} \binom{n}{k} p^k (1-p)^{n-k}$$
$$= (p + (1-p))^n = 1,$$

since there are $\binom{n}{k}$ outcomes with exactly k 1s, and the second identity follows from Exercise 1.5.9.

What is the probability that the first 1 appears at the kth flip of the coin? We can compute this probability using the combinatorial ideas from Example 1.2.10. The event in question can be written as

$$A_k = \{\omega \in \Omega : \omega_1 = \cdots = \omega_{k-1} = 0, \omega_k = 1\}.$$

We can rewrite this in more familiar terms by defining the event B_i as the event that $\omega_i = 0$. With this definition, we can write

$$A_k = B_1 \cap B_2 \cap \cdots \cap B_{k-1} \cap B_k^c.$$

To compute the probability of A_k, we need to distinguish between members of A_k according to the number of 1s among the last $n - k$ positions. There are $\binom{n-k}{i}$ ways to put i 1s among the last $n - k$ positions, and any element in A_k that has i 1s among the last $n - k$ positions has probability $p(1-p)^{k-1} \times p^i(1-p)^{n-k-i}$. Therefore, we find that

$$P(A_k) = \sum_{i=0}^{n-k} \binom{n-k}{i} p(1-p)^{k-1} p^i (1-p)^{n-k-i}$$
$$= p(1-p)^{k-1} \sum_{i=0}^{n-k} \binom{n-k}{i} p^i (1-p)^{n-k-i} = p(1-p)^{k-1},$$

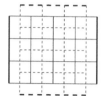 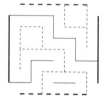

Figure 1.1: The first picture shows a network (solid lines) with its dual network (dashed lines). The second picture shows a realisation of a random network, together with the associated realisation in its dual. Note that in the solid line network, there is a connection from left to right, while there is no top to bottom connection in the dual network.

according again to Exercise 1.5.9.

However, there is a better way to compute $P(A_k)$. Note that we have constructed the experiment in such a way that the events B_i are independent. Indeed, we built our probability measure in such a way that any outcome with k 1s and $n - k$ 0s has probability $p^k(1 - p)^{n-k}$, which is the product of the individual probabilities. Hence we see that

$$
\begin{aligned}
P(A_k) &= P(B_1 \cap B_2 \cap \cdots \cap B_{k-1} \cap B_k^c) \\
&= P(B_1)P(B_2) \cdots P(B_{k-1})P(B_k^c) = (1 - p)^{k-1}p.
\end{aligned}
$$

$\square$

Example 1.5.11 (Random networks). The theory that we discussed so far can sometimes be used in a very surprising way. In this example we consider random networks. Consider a $(n + 1) \times n$ rectangle, subdivided into $n(n + 1)$ squares. See the solid line network in the first picture in Figure 1.5, where $n = 3$. Suppose now that for each side of a square, we flip a fair coin. When head comes up, we delete the line segment, and when tail comes up we leave it there. We do not flip coins for vertical edges on the boundary of the rectangle, that is why these are drawn with a different thickness. What results is a so-called random network and we can explore its properties. For instance, does the resulting network allow for a connection from the left side of the rectangle to the right side? (Now you understand why we do not flip coins for the vertical edges on the boundary; for the existence or non-existence of a connection from left to right, these edges do not matter.) It turns out that the probability of the event that such a connection exists does not depend on n. This is quite remarkable. Even more remarkable is that we can prove this with the little theory we have had so far. In order to state the result, we introduce some notation. As usual, we denote the sample space by Ω. An element of Ω can be written as $\omega = (\omega(e_1), \omega(e_2), \ldots, \omega(e_k))$, where the e_i's denote the edges of the network, and $\omega(e_i)$ is equal to 1 when the edge is present in ω, and equal to 0 otherwise. The states of different edges are independent, that is, the events $\omega(e_j) = 1$ form an independent collection of events. Since each edge has probability $1/2$ of being present, each possible outcome has the same probability $|\Omega|^{-1}$. We denote this probability measure by P_n. The event that the resulting network has a connection from left to right is denoted by $LR(n)$.

Theorem 1.5.12. *It is the case that*

$$P_n\left(LR(n)\right) = \frac{1}{2},$$

for all n.

Proof. Since every outcome has the same probability, it is enough to show that the number of outcomes in which there is a connection from left to right, is the same as the number of outcomes for which there is no such connection. We will show this by constructing a one-to-one correspondence between outcomes with and without such a connection. To do this we introduce the notion of a *dual network*.

To each network of size $(n+1) \times n$ we associate a so called dual network. This is drawn in dashed lines in Figure 1.1, and obtained as follows. As is clear from the picture, a dual network is rectangular of size $n \times (n+1)$. Now consider a certain outcome of the experiment, for instance the outcome in the second picture in Figure 1.1. With each realisation of the random network, we associate a realisation in the dual network as follows. Each edge of the network crosses exactly one edge of the dual network. Whenever an edge e of the original network is present in the outcome, the corresponding edge in the dual network – that is, the edge which crosses e – is taken out. Otherwise, the edge remains. Now note a most remarkable property. If the original network has a connection from left to right, then the dual network has no connection from top to bottom. If, on the other hand, the original network has no left-right connection, then there is a top to bottom connection in the dual.

With this last observation, our one-to-one correspondence becomes almost obvious: to each outcome of the original network which has a left to right connection, we associate the corresponding outcome in the dual and then rotate it by 90 degrees. This gives the desired one-to-one correspondence, and finishes the proof. □

Example 1.5.13. In this example, we introduce a very important probability mass function on the sample space $\mathbb{N}$. We let, for any $\lambda > 0$, P_λ be the probability mass function defined by

$$P_\lambda(k) = e^{-\lambda}\frac{\lambda^k}{k!}, \tag{1.1}$$

for $k = 0, 1, \ldots$

♠ **Exercise 1.5.14.** Prove that this is indeed a probability measure. You will need the *exponential series* for this:

$$e^\lambda = \sum_{k=0}^{\infty} \frac{\lambda^k}{k!}.$$

We shall now explain why this probability mass function is introduced in a section devoted to independence. You should interpret this explanation as intuitive motivation for the probability mass function in (1.1).

Suppose that we want to investigate arrivals of customers at a shop between time $t = 0$ and $t = 1$. As a first approximation, we could divide the unit time interval into n disjoint intervals of length $1/n$, and make the assumption that in each time interval of lenght $1/n$ at most one customer can arrive. Also, we make the assumption that the probability that a customer does in fact arrive in a given interval is proportional to the length of that interval. Hence, we assume that there is $\lambda > 0$ such that the probability that a customer arrives in an interval of length $1/n$ is equal to λ/n. Finally, we assume that arrival of customers in different time intervals are independent. With these assumptions we see that the probability that exactly k customers arrive in the interval from $t = 0$ to $t = 1$, should be the same as the probability that k heads come up when we throw n times with a coin which has probability λ/n of seeing head. That is, we have that this probability should be equal to

$$
\begin{aligned}
f_n(k) \quad &:= \quad \binom{n}{k} \left(\frac{\lambda}{n}\right)^k \left(1 - \frac{\lambda}{n}\right)^{n-k} \\
&= \quad \frac{n!}{k!(n-k)!} \left(\frac{\lambda}{n}\right)^k \left(1 - \frac{\lambda}{n}\right)^n \left(1 - \frac{\lambda}{n}\right)^{-k} \\
&= \quad \frac{\lambda^k}{k!} \left(1 - \frac{\lambda}{n}\right)^n \frac{n!}{n^k(n-k)!} \left(1 - \frac{\lambda}{n}\right)^{-k}.
\end{aligned}
$$

So far we have fixed n, but it is quite natural to make n larger and larger, so that the approximation becomes finer and finer – and hopefully also better and better. Therefore, it is natural to investigate what happens with the expression for $f_n(k)$ when we take the limit as $n \to \infty$. Using one of the standard limits from calculus, namely

$$
\lim_{n \to \infty} \left(1 - \frac{\lambda}{n}\right)^n = e^{-\lambda},
$$

we see that the first two terms converge to

$$
\frac{\lambda^k}{k!} e^{-\lambda},
$$

when $n \to \infty$. It is an exercise to prove that the last two terms both converge to 1 as $n \to \infty$, from which we conclude that

$$
\lim_{n \to \infty} f_n(k) = \frac{\lambda^k}{k!} e^k,
$$

which corresponds to the probability mass function in (1.1).   □

♠ **Exercise 1.5.15.** Show that the last two terms indeed converge to 1.

Example 1.5.16 (Island problem). This problem was responsible for an interesting debate in the probability literature. Consider an island with $n+2$ inhabitants. One of them is killed, and the murderer must be one of the inhabitants of the island. Police investigators discover a DNA profile at the scene of the crime. Scientists are able to say that this particular DNA profile occurs in a fraction p of all people. Now the police starts a big screening of all the inhabitants of the island. The first person to be screened, let's call him John Smith, turns out to have this particular DNA profile. What is the probability that John Smith is the murderer?

In order to say something about this, we need to turn the situation into a real mathematical model. We assign to every inhabitant of the island a DNA profile, and the probability that someone has the profile found at the scene of the crime is p. Different persons have independent profiles. Furthermore, we assume that each person is the murderer with probability $1/(n+1$.

Here are two 'solutions' to the problem:

(1) Apart from John Smith, there are n other people on the island which are the potential murderer. Suppose first that John Smith is the only one with the profile. In that case he is the murderer for sure. If there is exactly one other person with the profile, then the probability that John is the criminal is $1/2$. In general, if there are k other people with the profile on the island, then the probability that John Smith is our man, is equal to $1/(k+1)$. Writing A_k for the event that exactly k other people have the same profile as John Smith, we have

$$P(A_k) = \binom{n}{k} p^k (1-p)^{n-k}. \tag{1.2}$$

We then compute, writing G for the event that John Smith is the killer,

$$
\begin{aligned}
P(G) &= \sum_{k=0}^{n} P(G|A_k)P(A_k) \\
&= \sum_{k=0}^{n} \frac{1}{1+k} \binom{n}{k} p^k (1-p)^{n-k} \\
&= \frac{1}{p(n+1)} \sum_{k=0}^{n} \frac{(n+1)!}{(k+1)!(n-k)!} p^{k+1}(1-p)^{(n+1)-(k+1)} \\
&= \frac{1}{p(n+1)} \sum_{k=1}^{n+1} \binom{n+1}{k} p^k (1-p)^{(n+1)-k} \\
&= \frac{1}{p(n+1)} \left(1 - (1-p)^{n+1}\right),
\end{aligned}
$$

where the last equality follows from Exercise 1.5.9.

(2) We can, alternatively, apply Bayes' rule, Theorem 1.4.13. Denote the event that John Smith is the murderer by G, and denote the event of finding John Smiths'

particular DNA profile at the scene of the crime by E. Before seeing this DNA profile, any inhabitant of the island is equally likely to be the murderer, and we therefore have $P(G) = 1/(n+1)$ and $P(G^c) = n/(n+1)$. We want to compute the probability of G after the DNA evidence E, that is, we want to compute $P(G|E)$.

If John Smith is the murderer, the event E occurs with probability 1, that is, $P(E|G) = 1$. If John Smith is not the murderer, then the real murderer has with probability p the same profile as John Smith, and therefore, $P(E|G^c) = p$. We can now, with this information, compute the probability that John Smith is the murderer, given the event that his DNA profile was found at the scene of the crime, that is, we can compute $P(G|E)$:

$$
\begin{aligned}
P(G|E) &= \frac{P(E|G)P(G)}{P(E|G)P(G) + P(E|G^c)P(G^c)} \\
&= \frac{1/(n+1)}{1/(n+1) + (pn)/(n+1)} \\
&= \frac{1}{1+pn}.
\end{aligned}
$$

This is alarming, since the two methods give different answers. Which one is correct? We will now explain why method (1) is wrong. In method (1), we said that the probability that there are k other people with John Smiths' profile is given by (1.2). This seems obvious, but is, in fact, not correct. The fact that the first person to be checked has the particular DNA profile, says something about the total number of individuals with this profile. The situation is very similar to the situation in Example 1.4.6. In that example, even when we know that a family has at least one boy, when we then actually see a boy opening the door, this new information does change the conditional probability that the family has two boys. The bare fact that a boy opened the door, makes it more likely that there are two boys. Similarly, the fact that the first person to be screened has the DNA profile, makes it more likely that there are more such persons.   □

♠ **Exercise 1.5.17.** Method (1) above can be made correct by taking into account the so called *size bias* which we tried to explain above. Can you compute the right answer via this method? This is not so easy.

1.6   A First Law of Large Numbers

One of the most basic intuitive ideas in probability theory is the idea that when we flip a fair coin very often, the fraction of heads should be roughly $1/2$. A so called *law of large numbers* is a mathematical formulation of this idea. We will encounter many laws of large numbers in this book.

It is perhaps surprising that with a minimum of technical machinery – after all we only count – we can already state and prove the first of these laws of large numbers. We return to the setting of Example 1.2.10. Consider the event B_k that we see exactly k tails in n coin flips. For our present purpose, the number n is not fixed and we shall take a limit for $n \to \infty$ in a moment. This means that we should be careful with our notation, and we better express the dependence on n of the various quantities. Hence we denote the probability measure by P_n, and the event of seeing exactly k tails by $B_{k,n}$. In this notation, recall that

$$P_n(B_{k,n}) = \binom{n}{k} 2^{-n}.$$

The idea of the law of large numbers is that when n is large, the fraction of tails in the outcome should be close to $1/2$. One way of expressing this is to say that the probability that this fraction is close to $1/2$ should be large. Therefore, we consider the event that after n coin flips, the fraction of tails is between $1/2 - \epsilon$ and $1/2 + \epsilon$. We can express this event in terms of the $B_{k,n}$'s by

$$\bigcup_{n(\frac{1}{2}-\epsilon) \leq k \leq n(\frac{1}{2}+\epsilon)} B_{k,n}.$$

Theorem 1.6.1 (Law of large numbers). *For any $\epsilon > 0$, we have*

$$P_n \left(\bigcup_{n(\frac{1}{2}-\epsilon) \leq k \leq n(\frac{1}{2}+\epsilon)} B_{k,n} \right) \to 1,$$

as $n \to \infty$.

Proof. We shall prove that

$$P_n \left(\bigcup_{n(\frac{1}{2}+\epsilon) < k \leq n} B_{k,n} \right) \to 0, \tag{1.3}$$

and

$$P_n \left(\bigcup_{0 \leq k < n(\frac{1}{2}-\epsilon)} B_{k,n} \right) \to 0, \tag{1.4}$$

as $n \to \infty$. This is enough, since $P_n(\cup_{k=0}^n B_{k,n})$ is equal to 1, and hence the probability of the union over all the remaining indices must converge to 1. We shall only prove (1.3), the proof of (1.4) is similar and left to you.

First observe that

$$P_n \left(\bigcup_{k > n(\frac{1}{2}+\epsilon)} B_{k,n} \right) = \sum_{k > n(\frac{1}{2}+\epsilon)} P_n(B_{k,n})$$

$$= \sum_{k > n(\frac{1}{2}+\epsilon)} \binom{n}{k} 2^{-n}.$$

The following surprising trick is quite standard in probability theory. For $\lambda > 0$, we find that

$$\sum_{k > n(\frac{1}{2}+\epsilon)} \binom{n}{k} 2^{-n} \leq \sum_{k > n(\frac{1}{2}+\epsilon)} e^{\lambda(k-n(\frac{1}{2}+\epsilon))} \binom{n}{k} 2^{-n}$$

$$= e^{-\lambda n \epsilon} \sum_{k > n(\frac{1}{2}+\epsilon)} \binom{n}{k} \left(\frac{1}{2}e^{\lambda/2}\right)^k \left(\frac{1}{2}e^{-\lambda/2}\right)^{n-k}$$

$$\leq e^{-\lambda n \epsilon} \sum_{k=0}^{n} \binom{n}{k} \left(\frac{1}{2}e^{\lambda/2}\right)^k \left(\frac{1}{2}e^{-\lambda/2}\right)^{n-k}$$

$$= e^{-\lambda n \epsilon} \left(\frac{1}{2}e^{\lambda/2} + \frac{1}{2}e^{-\lambda/2}\right)^n,$$

according to Proposition 1.5.9. It is not hard to show that for all $x \in \mathbb{R}$, $e^x \leq x + e^{x^2}$; see the forthcoming Exercise 1.6.2. Using this inequality, we find that the last expression is at most

$$e^{-\lambda n \epsilon} \left(e^{\lambda^2/4}\right)^n = e^{\lambda^2 n/4 - \lambda n \epsilon}.$$

Now we can find λ to minimise the right hand side, that is, $\lambda = 2\epsilon$. This then finally yields the bound

$$P_n \left(\bigcup_{k > n(\frac{1}{2}+\epsilon)} B_{k,n} \right) \leq e^{-\epsilon^2 n}, \tag{1.5}$$

which tends to zero when n tends to infinity. $\qquad\qquad\qquad\qquad\qquad\qquad\square$

♠ **Exercise 1.6.2.** Prove that for all $x \in \mathbb{R}$, we have $e^x \leq x + e^{x^2}$.

1.7 Exercises

Exercise 1.7.1. In this exercise, A and B are events. Show that
(a) $(A \cap B^c) \cup B = A \cup B$;
(b) $(A \cup B) \cap B^c = A\backslash B$;
(c) $(A \cup B) \cap (A \cap B)^c = (A\backslash B) \cup (B\backslash A)$;
(d) $A = (A \cap B) \cup (A \cap B^c)$;
(e) $(A \cap B) \cup (A \cap C) = A \cap (B \cup C)$;
(f) $(A \cup B) \cap (A \cup C) = A \cup (B \cap C)$.

Exercise 1.7.2. In this exercise, A, B and C are events. Express the following events in A, B and C:
(a) Only A occurs;

(b) A and B occur, but C does not occur;
(c) All three events occur;
(d) Exactly one of the three events occurs;
(e) At most one of the events occurs;
(f) None of the three events occurs;
(g) At least two of the three events occur.

Exercise 1.7.3. Suppose we throw a die twice. What is the probability that
(a) the two outcomes are the same?
(b) the two outcomes are different and sum up to 8?
(c) the sum of the outcomes is 10?

Exercise 1.7.4. What is the probability that all four queens end up at one player in a play of bridge?

Exercise 1.7.5. An urn contains 7 red and 5 blue balls, which we take out (without looking!), one by one.
(a) What is the probability that the first ball is blue?
(b) What is the probability that the last ball is blue?
(c) What is the probability that the last ball is blue, given that the first ball is blue?

Exercise 1.7.6. Consider four boxes, numbered 1 to 4. We throw four balls in the boxes in such a way that each ball ends up in any particular box with probability $1/4$, and in such a way that different balls are thrown independently.
(a) What is the probability that there will be at least one empty box?
(b) What is the probability that there is exactly one empty box?
(c) What is the probability that box 1 remains empty?

Exercise 1.7.7. Suppose there are three men and four women who have to be arranged in a circle. If we do this randomly, what is the probability that all men stand together in one group?

Exercise 1.7.8. Consider a group of four people. Everybody writes down the name of one other (random) member of the group. What is the probability that there is at least one pair of people who wrote down each others name?

Exercise 1.7.9. Suppose that we play bridge. Each player receives 13 cards. What is the probability that south receives exactly 8 spades, and north the remaining 5?

Exercise 1.7.10. Suppose we dial a random number on my telephone, the number is six digits long. What is the probability that
(a) the number does not contain a 6;
(b) the number contains only even digits;
(c) the number contains the pattern 2345;
(d) the number contains the pattern 2222.

Exercise 1.7.11. Suppose that we order the numbers $1, 2, \ldots, n$ completely randomly. What is the probability that 1 is immediately followed by 2?

Exercise 1.7.12. We choose an integer N at random from $\{1, 2, \ldots, 10^3\}$. What is the probability that N is divisible by 3? by 5? by 105? How would your answer change if 10^3 is replaced by 10^k, as k gets larger and larger?

Exercise 1.7.13. Consider the complete graph K_4 with four vertices; all vertices are connected by an edge to all other vertices. Suppose now that we flip an unbiased coin for each edge. If heads comes up, we leave the edge where it is, if tails comes up we remove the edge.
(a) What is the probability that two given vertices are still connected after the removal of the edges?
(b) What is the probability that the graph remains connected?
(c) What is the probability that a given vertex becomes isolated?

Exercise 1.7.14. Suppose that we have a tennis tournament with 32 players. Players are matched in a completely random fashion, and we assume that each player always has probability $1/2$ to win a match. What is the probability that two given players meet each other during the tournament.

Exercise 1.7.15. Peter and Paul have a disagreement, and they want to make a decision by throwing a coin. Paul suspects that the coin is biased. Design a rule so that they can come to a fair decision.

Exercise 1.7.16. Consider n pair of shoes. Suppose that we take $2r$ of these (without looking of course), where $2r < n$. What is the probability that there is no pair among these $2r$ shoes? Can you also compute the probability that among these $2r$ shoes, there is exactly one pair?

Exercise 1.7.17. Consider two urns, one containing 5 red and 10 white balls, and the other with 5 white and 10 red balls. Now choose one of the urns randomly, and take two random balls from the chosen urn. Let A be the event that the first ball is red, and B be the event that the second ball is white. Are A and B independent?

Exercise 1.7.18. We draw a ball from an urn with 3 red and 2 blue balls. If the ball is red, we draw a second ball from another urn containing 4 red and 1 blue ball. If the first ball is blue, we draw a second ball from an urn with just 4 blue balls.
(a) What is the conditional probability that the second ball is red, given that the first ball is red?
(b) What is the probability that both balls are red?
(c) What is the probability that the second ball is blue?

Exercise 1.7.19. Suppose that we send a message using some coding so that only 0's and 1's are sent. On average, the ratio between the number of 0's and 1's that are sent is $3/4$. As a result of problems with the connection, each 0 sent is received as a 1 with probability $\frac{1}{4}$, and each 1 sent is received as a 0 with probability $\frac{1}{3}$. Compute the probability that a symbol received as a 1, was also sent as a 1.

Exercise 1.7.20. Three people, A, B and C play a game in which they throw coins, one after the other. A starts, then B, then C, then A again, etcetera. The person who throws heads first, wins the game. Construct an appropriate sample space for this game, and find the probability that A wins the game.

Exercise 1.7.21. Suppose that 20 rabbits live in a certain region. We catch 5 of them, mark these, and let them go again. After a while we catch 4 rabbits. Compute the probability that exactly 2 of these 4 are marked. Be very precise about the assumptions that you make when you compute this probability.

Exercise 1.7.22. Suppose you want to ask a large number of people a question to which the answer 'yes' is embarrassing. As the question is asked, a coin is tossed, out of sight of the questioner. If the answer would have been 'no' and the coin shows head, then the answer 'yes' is given. Otherwise people respond truthfully. Do you think that this is a good procedure?

Exercise 1.7.23. A fair coin is tossed n times. Let H_n and T_n denote the number of heads and tails among these n tosses. Show that, for any $\epsilon > 0$, we have

$$P_n \left(-\epsilon \le \frac{H_n - T_n}{n} \le \epsilon \right) \to 1,$$

as $n \to \infty$.

Exercise 1.7.24 (Simpson's paradox). A researcher wants to determine the relative efficacies of two drugs. The results (differentiated between men and women) were as follows.

women	drug I	drug II
succes	200	10
failure	1800	190

men	drug I	drug II
succes	19	1000
failure	1	1000

We are now faced with the question which drug is better. Here are two possible answers:
(1) Drug I was given to 2020 people, of whom 219 were cured. Drug II was given to 2200 people, of whom 1010 were cured. Therefore, drug II is much better,
(2) Amongst women, the succes rate of drug I is $1/10$, and for drug II the succes rate is $1/20$. Amongst men, these rates are $19/20$ and $1/2$ respectively. In both cases, that is, for both men and women, drug I wins, and is therefore better.
    Which of the two answers do you believe? Can you explain the paradox?

Exercise 1.7.25. Suppose that we want to distribute five numbered balls over three boxes I, II and III. Each ball is put in a random box, independently of the other balls. Describe an appropriate sample space and probability measure for this

experiment. Compute the probability that
(a) box I remains empty;
(b) at most one box remains empty;
(c) box I and II remain empty.

Exercise 1.7.26. An urn contains 10 white, 5 yellow and 10 black balls. We pick a random ball. What is the probability that the ball is yellow, given that it is not black?

Exercise 1.7.27. Peter and Paul both have 24 photos from their holidays. Suddenly the wind blows them all away. After an extensive search, they manage to find a total of 40 photos back.
(a) What is the probability that Peter's photos have all been found?
(b) Suppos they find the photos back one by one. What is the probability that they find all photos of Peter before a single phote of Paul has been found?
(c) What is the probability that there are more photos found from Peter than from Paul?

Exercise 1.7.28. We throw a fair die twice. A is the event that the sum of the throws is equal to 4, B is the event that at least one of the throws is a 3. Are A and B independent?

Exercise 1.7.29. We take two cards from a regular deck of 52 cards, withour replacement. A is the event that the first card is a spade, B is the event that the second card is a spade. Are A and B independent?

Exercise 1.7.30. We choose a month of the year so that each month has the same probability. Let A be the event that we choose an 'even' months (that is, februari, april, ...) and let B be the event that the outcome is in the first half of the year. Are A and B independent? If C is the event that the outcome is a summer month (that is, june, july, august), are A and C independent?

Exercise 1.7.31. It is known that 5% of the men is colour blind, and $\frac{1}{4}$% of the women is colour blind. Suppose that there are as many men as women. We choose a person, which turns out to be colour blind. What is the probability that this person is a man?

Exercise 1.7.32. Suppose that we have a very special die, namely with exactly k faces, where k is a prime. The faces of the die are numbered $1, \ldots, k$. We throw the die and see which number comes up.
(a) What would be an appropriate sample space and probability measure?
(b) Suppose that the events A and B are independent. Show that A or B can only be the full sample space, or the empty set.

Exercise 1.7.33. Consider a family which we know has two children. I tell you that it is not the case that they have two girls. What is the (conditional) probability that they have two boys?

Now suppose that I suddenly see the father of this family walking with a little boy holding his hand. Does this change the conditional probability that the family has two boys?

Exercise 1.7.34. Suppose we throw a fair coin n times. Let A be the event that we see at most one head, and let B be the event that we see at least one head and at least one tail. Show that A and B are independent when $n = 3$ and show that they are not independent when $n \neq 3$.

Exercise 1.7.35. Suppose that I have two coins in my pocket. One ordinary, fair coin, and one strange coin with heads on either side. I pick a random coin out of my pocket, throw it, and it turns out that head comes up.
(a) What is the probability that I have thrown the fair coin?
(b) If I throw the same coin again, and head comes up again, what is the probability that I have thrown the fair coin?

Exercise 1.7.36. A pack contains m cards, labelled $1, 2, \ldots, m$. The cards are dealt out in a random order, one by one. Given that the label of the kth card dealt is the largest of the first k cards, what is the probability that it is also the largest in the whole pack?

Exercise 1.7.37 (de Méré's paradox). Which of the following two events has the highest probability:
(1) at least one 6, when we throw a die 4 times;
(2) at least one double 6, when we throw two dice 24 times.

Exercise 1.7.38. Let $A_1, A_2, \ldots$ be events. Show that

$$P\left(\bigcap_{i=1}^{n} A_i\right) \geq \sum_{i=1}^{n} P(A_i) - (n-1),$$

for all $n = 1, 2, \ldots$

Exercise 1.7.39. Let $A_1, A_2, \ldots, A_n$ be events such that $P(A_1 \cap \cdots \cap A_{n-1}) > 0$. Prove that

$$P\left(\bigcap_{i=1}^{n} A_i\right) = P(A_1)P(A_2|A_1)P(A_3|A_1 \cap A_2) \cdots$$
$$\cdots P(A_n|A_1 \cap A_2 \cap \cdots \cap A_{n-1}).$$

Exercise 1.7.40. Consider the following game: player I flips a fair coin $n+1$ times; player II flips a fair coin n times. Show that the probability that player I has more heads than player B, is equal to $\frac{1}{2}$. Is this counterintuitive, given the fact that player I flips the coin one extra time?

Exercise 1.7.41. Consider two events A and B, both with positive probability. When is it the case that $P(A|B) = P(B|A)$?

Chapter 2

Random Variables and Random Vectors

It often happens that we do not really care about the outcome of an experiment itself, but rather we are interested in some consequence of this outcome. For instance, a gambler is not primarily interested in the question whether or not heads comes up, but instead in the financial consequences of such an outcome. Hence the gambler is interested in a *function* of the outcome, rather than in the outcome itself. Such a function is called a *random variable* and in this chapter we define and study such random variables.

2.1 Random Variables

Suppose that we flip a coin n times. Each time tails comes up we lose one euro, and when heads comes up we win one euro. The sample space corresponding to this experiment could be $\{-1, 1\}^n$, the set of sequences of -1s and 1s of length n. How can we express our fortune after n flips?

Writing $\omega = (\omega_1, \ldots, \omega_n)$ as usual, where $\omega_i = 1$ if the ith flip is a head, my winnings after n flips are equal to

$$S(\omega) = \sum_{i=1}^{n} \omega_i.$$

Thus our winnings is a mapping $S : \Omega \to \mathbb{R}$. Such mappings are called *random variables*.

Definition 2.1.1. A *random variable* X is a mapping from a sample space Ω into $\mathbb{R}$.

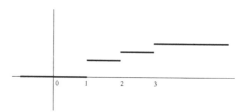

Figure 2.1: A sketch of the distribution function of X.

Typically, we denote random variables with letters near the end of the alphabet, like X, Y and Z. We are often interested in the probability that a random variable takes certain values. Hence we are interested in

$$P(\{\omega : X(\omega) = x\}),$$

for all appropriate x. We shall often write $\{X = x\}$ for $\{\omega : X(\omega) = x\}$ and $P(X = x)$ for $P(\{\omega : X(\omega) = x\})$.

Definition 2.1.2. The *probability mass function* of a random variable X is the function $p_X : \mathbb{R} \to [0, 1]$ given by

$$p_X(x) = P(X = x).$$

Definition 2.1.3. The *distribution function* of a random variable X is the function $F_X : \mathbb{R} \to [0, 1]$ given by
$$F_X(x) = P(X \leq x).$$

Example 2.1.4. Suppose that the random variable X takes the value 1 with probability 1, that is, $P(X = 1) = 1$. The distribution function is then given by $F_X(x) = 1$, for $x \geq 1$, and $F_X(x) = 0$ for $x < 1$. Note that this function is continuous from the right but not continuous from the left at $x = 1$.   □

Example 2.1.5. Consider a random variable X with $P(X = 1) = 1/2$, $P(X = 2) = 1/4$ and $P(X = 3) = 1/4$. The distribution function of X is given by

$$F_X(x) = \begin{cases} 0 & \text{if } x < 1, \\ 1/2 & \text{if } 1 \leq x < 2, \\ 3/4 & \text{if } 2 \leq x < 3, \\ 1 & \text{if } x \geq 3, \end{cases}$$

see Figure 2.1.   □

Example 2.1.6. Suppose that we toss a fair coin n times. The number of heads is a random variable which we denote by X. The probability mass function of X is given by

$$p_X(k) = \binom{n}{k} 2^{-n},$$

for $k = 0, \ldots, n$, and $p_X(k) = 0$ for all other values of k. Hence its distribution function is given by

$$F_X(x) = 2^{-n} \sum_{0 \le k \le x} \binom{n}{k},$$

for $0 \le x \le n$; $F_X(x) = 0$ for $x < 0$; $F_X(x) = 1$ for $x > n$. $\qquad\square$

Example 2.1.7 (Binomial distribution). A random variable X is said to have a *binomial distribution* with parameters $n \in \mathbb{N}$ and $p \in [0, 1]$ if

$$P(X = k) = \binom{n}{k} p^k (1 - p)^{n-k},$$

for $k = 0, 1, \ldots, n$. We have seen examples of such random variables in Chapter 1 when we discussed coin flips. The random variable X represents the number of heads when we flip a coin n times, where each flip gives heads with probability p. In particular, we can write such a random variable X as a sum

$$X = \sum_{i=1}^{n} Y_i,$$

where $Y_i = 1$ if the ith flip yields a head, and $Y_i = 0$ otherwise. $\qquad\square$

Example 2.1.8 (Poisson distribution). A random variable X is said to have a *Poisson distribution* with parameter $\lambda > 0$ if

$$P(X = k) = \frac{\lambda^k}{k!} e^{-\lambda},$$

for $k = 0, 1, \ldots$ We have come across this distribution already in Example 1.5.13, where it naturally appeared as a limit of binomial random variables. $\qquad\square$

Example 2.1.9 (Geometric distribution). A random variable X has a *geometric distribution* with parameter $p \in (0, 1]$ if

$$P(X = k) = p(1 - p)^{k-1},$$

for $k = 1, 2, \ldots$ We have seen a random variable with this distribution in Example 1.5.10. We can view X as representing the waiting time until the first time heads comes up in a sequence of independent coin flips.

There is one subtle thing that needs attention at this point. We interpreted a geometric random variable as the waiting time for the first head to come up in a sequence of coin flips. This suggests that we want to define X on a sample space which corresponds to infinitely many coin flips. Indeed, the first head may come up at any time: there is no bound on the time at which the first head comes up. However, all our sample spaces so far contained at most countably many points, and we shall see later that any sample space which represents infinitely many coin flips is necessarily uncountable. Is this a problem? In some sense yes, but in another sense no. If we flip a coin, and we are only interested in the time that the first head comes up, then it is enough to take $\{1, 2, \ldots\} \cup \{\infty\}$ as our sample space. An outcome k then corresponds to the event that the first head appears at the kth flip, and the outcome ∞ corresponds to the event that we never see a head. This sample space clearly does not capture all events in the coin flips. For instance, an outcome $X = 6$ says nothing about the 7th, 8th,... coin flip. But if we only want to know something about the time of the first head, then this is no problem. Hence, the choice of the sample space depends on the question or object you want to study. When it comes down to defining an appropriate probability measure for this experiment, we have seen that it is reasonable to take $P(k) = (1 - p)^{k-1}p$, for $k = 1, 2, \ldots$ What should $P(\infty)$ be? Well, since $\sum_{k=1}^{\infty} P(k) = 1$, there is no probability mass left for ∞, and we conclude, *a fortiori*, that $P(\infty) = 0$. That is, in this model it is certain that we will throw a head at some point. This is of course in complete agreement with our probabilistic intuition.   □

♠ **Exercise 2.1.10.** Make sure that you understand why $p = 0$ is excluded for the geometric distribution.

Example 2.1.11 (Negative binomial distribution). The random variable X has a *negative binomial distribution* with parameters $r \in \mathbb{N}$ and $p \in (0, 1]$ if

$$P(X = k) = \binom{k - 1}{r - 1} p^r (1 - p)^{k-r},$$

for $k = r, r + 1, \ldots$ This looks rather complicated. However, we can interpret X as the waiting time until head comes up for the rth time in a sequence of coin flips. To see this, note that in order to see heads for the rth time at the kth flip, we must have seen exactly $r - 1$ heads among the first $k - 1$ flips. This can be realised in $\binom{k-1}{r-1}$ ways. The probability that this happens for a particular realisation with $r-1$ heads among the first $k-1$ flips is simply $p^{r-1}(1-p)^{(k-1)-(r-1)} = p^{r-1}(1-p)^{k-r}$. The requirement that the kth flip is also a head gives an extra factor p, and then we arrive at the given formula.

Observe that we can split the waiting time until the rth head into r pieces, each piece consisting of the waiting time until the next heads appears. This makes it clear that a random variable X with a negative binomial distribution with parameters r and p, can be seen as the sum of r geometrically distributed random variables.   □

♠ **Exercise 2.1.12.** Show that the probabilities of the negative binomial distribution sum up to 1.

It is useful to list a number of properties shared by all distribution functions.

Theorem 2.1.13. *A distribution function F of a random variable X has the following properties:*

(a) $\lim_{x \to \infty} F(x) = 1$;

(b) $\lim_{x \to -\infty} F(x) = 0$;

(c) F *is non-decreasing*;

(d) F *is continuous from the right*;

(e) $P(X > x) = 1 - F(x)$;

(f) $P(x < X \le y) = F(y) - F(x)$;

(g) $P(X = x) = F(x) - \lim_{y \uparrow x} F(y)$.

In the proof of this theorem, we need the following lemma, which is of independent interest.

Lemma 2.1.14. (a) *Let* $A_1 \subseteq A_2 \subseteq A_3 \subseteq \cdots$, *and let* $A = \bigcup_{i=1}^{\infty} A_i$. *Then*

$$P(A) = \lim_{i \to \infty} P(A_i).$$

(b) *Let* $B_1 \supseteq B_2 \supseteq B_3 \supseteq \cdots$, *and let* $B = \bigcap_{i=1}^{\infty} B_i$. *Then*

$$P(B) = \lim_{i \to \infty} P(B_i).$$

Proof. (a) We can write $A = A_1 \cup (A_2 \backslash A_1) \cup (A_3 \backslash A_2) \cup \cdots$ as a union of pairwise disjoint events. Hence,

$$
\begin{aligned}
P(A) &= P(A_1) + \sum_{i=1}^{\infty} P(A_{i+1} \backslash A_i) \\
&= P(A_1) + \lim_{n \to \infty} \sum_{i=1}^{n-1} (P(A_{i+1}) - P(A_i)) \\
&= \lim_{n \to \infty} \{P(A_1) + P(A_2) - P(A_1) + P(A_3) - P(A_2) + \cdots \\
&\quad + P(A_n) - P(A_{n-1})\} = \lim_{n \to \infty} P(A_n).
\end{aligned}
$$

(b) See Exercise 2.1.15. $\qquad \square$

♦ **Exercise 2.1.15.** Prove Lemma 2.1.14(b).

Proof of Theorem 2.1.13. For (a), let A_i be the event that $X \le i$. Then, as $i \to \infty$, $A_i \uparrow \Omega$, and the statement follows from Lemma 2.1.14(a). (b) is proved similarly and left as an exercise. To prove (c), note that when $x < y$, $\{X \le x\} \subseteq \{X \le y\}$. The statement now follows from Lemma 1.3.1(c).

To prove (d), note that

$$\{X \le x\} = \bigcap_{n=1}^{\infty} \left\{X \le x + \frac{1}{n}\right\},$$

It then follows from Lemma 2.1.14(b) that

$$F(x) \;=\; P\left(X \le x\right) = \lim_{n \to \infty} P\left(X \le x + \frac{1}{n}\right)$$

$$=\; \lim_{n \to \infty} F(x + \frac{1}{n}).$$

Since F is a monotone function, $\lim_{h \downarrow 0} F(x + h)$ exists and by the previous computation, this limit must be $F(x)$.

The statement in (e) follows from Lemma 1.3.1(b). The proof of (f) is left as an exercise.

For (g), observe that

$$\{X = x\} = \bigcap_{n=1}^{\infty} \{x - 1/n < X \le x\}.$$

Hence, we have according to Lemma 2.1.14(b) that

$$P(X = x) \;=\; \lim_{n \to \infty} P(x - 1/n < X \le x)$$

$$=\; \lim_{n \to \infty} \left(F(x) - F(x - 1/n)\right)$$

$$=\; F(x) - \lim_{n \to \infty} F(x - 1/n).\qquad \square$$

♠ **Exercise 2.1.16.** Complete the proof of Theorem 2.1.13.

♠ **Exercise 2.1.17.** Suppose that X has distribution function F. What is the distribution function of the random variable Y defined by $Y = aX + b$?

We end this section with a statement to the effect that the probability mass function of a random variable is uniquely determined by its distribution function, and vice versa.

Theorem 2.1.18. *Two random variables have the same probability mass function if and only if they have the same distribution function.*

Proof. Let X and Y be such that $p_X(x) = p_Y(x)$ for all x. Then

$$F_X(x) \;=\; P(X \le x) = \sum_{y:y \le x} p_X(y)$$

$$=\; \sum_{y:y \le x} p_Y(y) = F_Y(x).$$

Conversely, suppose that $F_X(x) = F_Y(x)$ for all x. Then it follows immediately from Theorem 2.1.13(g) that $p_X(x) = p_Y(x)$ for all x. $\square$

Definition 2.1.19. If two random variables X and Y have the same probability mass function or, equivalently, have the same distribution function, then we say that X and Y have the same *distribution*. Asking for the *distribution* of a random variable is asking for either its probability mass function, or its distribution function.

2.2   Independence

I have heard it said that probability theory is just analysis plus the concept of independence. I am not sure whether this makes any sense, but it certainly expresses the idea that independence plays a crucial role in probability theory.

We want two random variables to be called independent, if knowledge of the outcome of the first has no effect on the distribution of the second. Here is the formal definition.

Definition 2.2.1. The random variables $X_1, X_2, \ldots, X_n$ are called independent if the events $\{X_1 = x_1\}, \ldots, \{X_n = x_n\}$ are independent for any choice of $x_1, \ldots, x_n$.

The concept is best illustrated with some examples.

Example 2.2.2 (General coin tossing). Consider the general coin tossing in Example 1.5.10. Let $X(\omega) = \omega_1$ and $Y(\omega) = \omega_2$. That is, X is the outcome of the first coin flip, and Y is the outcome of the second. We already showed in Chapter 1 that X and Y are independent.   □

Example 2.2.3. Suppose we flip a fair coin 5 times. The random variable X takes the value 0 if the number of tails is even, and takes the value 1 if the number of tails is odd. The random variable Y takes the value 1 if the first throw gives tail, and 0 otherwise. A little thought (or a computation) reveals that $P(X = 0) = P(X = 1) = 1/2$ and of course we also have $P(Y = 0) = P(Y = 1) = 1/2$. The event $\{X = 0, Y = 1\}$ is the event that the first throw gives tail, and the total number of tails is odd. This last requirement means that the number of tails among the last four throws must be odd, and the probability that this happens is again $1/2$. Hence, we find that $P(X = 0, Y = 1) = 1/2 \times 1/2 = P(X = 0)P(Y = 1)$. A similar argument is valid for the other outcomes, and we conclude that X and Y are independent random variables.   □

♠ **Exercise 2.2.4.** How can you convince somebody that X and Y in the last example are independent, without doing any computation?

Example 2.2.5. Suppose we flip a coin a random number N of times, where

$$P(N = n) = e^{-\lambda}\frac{\lambda^n}{n!}, \quad n = 0, 1, 2, \ldots,$$

that is, N has a Poisson distribution. The flips are independent, and the probability of seeing heads is equal to p. Hence if we know that $N = n$, the number of heads is simply the number of heads in n coin flips with success probability p. We write X for the total number of heads, and Y for the total number of tails, so $X + Y = N$. The above conclusion can now be formalised as follows:

$$P(X = x | N = n) = \binom{n}{x}p^x(1 - p)^{n-x}.$$

Are X and Y independent? It is perhaps surprising that they are. To see this, we compute

$$
\begin{aligned}
P(X = x, Y = y) &= P(X = x, Y = y | N = x + y) P(N = x + y) \\
&= \binom{x + y}{x} p^x (1 - p)^y \frac{\lambda^{x+y}}{(x + y)!} e^{-\lambda} \\
&= \frac{(\lambda p)^x (\lambda(1 - p))^y}{x! y!} e^{-\lambda}.
\end{aligned}
$$

On the other hand, we have

$$
\begin{aligned}
P(X = x) &= \sum_{n \geq x} P(X = x | N = n) P(N = n) \\
&= \sum_{n \geq x} \binom{n}{x} p^x (1 - p)^{n-x} \frac{\lambda^n}{n!} e^{-\lambda} \\
&= \frac{(\lambda p)^x}{x!} e^{-\lambda p},
\end{aligned}
$$

and a similar computation gives

$$
P(Y = y) = \frac{(\lambda(1 - p))^y}{y!} e^{-\lambda(1-p)}.
$$

Hence, $P(X = x, Y = y) = P(X = x)P(Y = y)$, and we conclude that X and Y are independent. This is a most remarkable result. The point is that the number of times we flip the coin is itself random. □

When X and Y are independent, much more is true than what is stated in the definition of independence.

Theorem 2.2.6. *Let X and Y be two independent random variables. Then we have, for all $A, B \subset \mathbb{R}$,*

$$
P(X \in A, Y \in B) = P(X \in A)P(X \in B).
$$

Proof. We simply write

$$
\begin{aligned}
P(X \in A, Y \in B) &= \sum_{a \in A} \sum_{b \in B} P(X = a, Y = b) \\
&= \sum_{a \in A} \sum_{b \in B} P(X = a)P(Y = b) \\
&= \sum_{a \in A} P(X = a) \sum_{b \in B} P(Y = b) \\
&= P(X \in A)P(Y \in B).
\end{aligned}
$$

□

♠ **Exercise 2.2.7.** Show that X and Y are independent if and only if for all x, y,

$$P(X \leq x, Y \leq y) = P(X \leq x)P(Y \leq y).$$

When the random variables $X_1, X_2, \ldots, X_n$ are independent, and we consider functions of these random variables, these functions are also independent. For instance, X_1^2 and $X_2 - 7$ are also independent. This is articulated in the following theorem.

Theorem 2.2.8. *Let $X_1, X_2, \ldots, X_n$ be independent random variables, and let, for $i = 1, \ldots, n$, g_i be a function $g_i : \mathbb{R} \to \mathbb{R}$. Then the random variables $g_1(X_1), g_2(X_2), \ldots, g_n(X_n)$ are also independent.*

Proof. We give the proof for the case $n = 2$, the full proof being asked for in the next exercise. We write

$$
\begin{aligned}
P(g_1(X_1) = a_1, g_2(X_2) = a_2) &= \sum_{(x_1, x_2): g_i(x_i) = a_i} P(X_1 = x_1, X_2 = x_2) \\
&= \sum_{(x_1, x_2): g_i(x_i) = a_i} P(X_1 = x_1)P(X_2 = x_2) \\
&= \sum_{x_1: g_1(x_1) = a_1} P(X_1 = x_1) \times \\
&\quad \times \sum_{x_2: g_2(x_2) = a_2} P(X_2 = x_2) \\
&= P(g_1(X_1) = a_1)P(g_2(X_2) = a_2).
\end{aligned}
$$
□

♠ **Exercise 2.2.9.** Give the full proof of the last result.

♠ **Exercise 2.2.10.** Let X_1, X_2 and X_3 be independent random variables, and let $g : \mathbb{R}^2 \to \mathbb{R}$ and $h : \mathbb{R} \to \mathbb{R}$. Show that $g(X_1, X_2)$ and $h(X_3)$ are independent random variables. Generalise this result.

2.3 Expectation and Variance

The concept of expectation refers to the 'average value' taken by a random variable. The expectation of a random variable will be defined as a sum, possibly with infinitely many terms, and it pays to spend a few lines about such sums.

Let $a_1, a_2, \ldots$ be real numbers. We only want to define $\sum_{n=1}^{\infty} a_n$ when this sum does not change when we change the order of the a_n's. A classical result from calculus tells us that this is the case when $S^+ = \sum_{n: a_n > 0} a_n$ and $S^- = -\sum_{n: a_n < 0} a_n$ are not both infinite with opposite sign. If S^+ and S^- are not both infinite with opposite sign, then we say that the sum $\sum_{n=1}^{\infty} a_n$ is *well defined*. Finally, we agree from now on that $\infty - \infty$ is not defined.

Definition 2.3.1. The *expectation* of a random variable X is given by

$$E(X) = \sum_x xP(X = x),$$

whenever this sum is well defined.

Why is this definition reasonable? At the beginning of this section, we said that the expectation refers to the average value taken by a random variable. we will now explain why this is the case with the above definition.

Let $x_1, \ldots, x_k$ be the outcomes of k independent random variables with the same distribution as some random variable X, and let, for each m, k_m be the number of x_i's which take the value m. Then we have that $\sum_{i=1}^{k} x_i = \sum_m m k_m$ and therefore the average $\frac{1}{k} \sum_{i=1}^{k} x_i$ of the x_i's is equal to $\sum_m m \frac{k_m}{k}$. As already mentioned in the previous chapter, the quotient k_m/k should, for large k, be close to $P(X = m)$, and the average of the x_i's should therefore be close to $\sum_m mP(X = m)$. This is precisely the definition of the expectation of X.

Example 2.3.2. Let X be a random variable with $P(X = 0) = 1/2$, and $P(X = 1) = 1/2$. Then $E(X) = 1/2$. This shows that the expectation of a random variable X need not be a value which can be taken by X. $\qquad\square$

Example 2.3.3. Suppose that X takes values -1, 1 and 2, with equal probability, and consider the random variable $Y = X^2$. Clearly, Y takes values 1 and 4 with probability $\frac{2}{3}$ and $\frac{1}{3}$ respectively. Therefore $E(Y) = 1 \cdot \frac{2}{3} + 4 \cdot \frac{1}{3} = 2$. $\qquad\square$

Example 2.3.4 (Poisson distribution). Recall that a random variable X is said to have a Poisson distribution with parameter $\lambda > 0$ if

$$P(X = k) = \frac{\lambda^k}{k!} e^{-\lambda},$$

for $k = 0, 1, \ldots$ To compute its expectation, we write

$$\begin{aligned}
E(X) &= \sum_{k=1}^{\infty} k \frac{\lambda^k}{k!} e^{-\lambda} \\
&= \lambda e^{-\lambda} \sum_{k=1}^{\infty} \frac{\lambda^{k-1}}{(k-1)!} \\
&= \lambda e^{-\lambda} e^{\lambda} = \lambda.
\end{aligned}$$

$\qquad\square$

Here follows an example with an infinite expectation.

Example 2.3.5 (St. Petersburg paradox). Suppose that you go to a casino, and that you can play the following game. A random number X is chosen in the casino, in such a way that $P(X = 2^n) = 2^{-n}$, for $n = 1, 2, \ldots$. The player receives this amount X from the casino. Of course, in order to play this game, you need to pay beforehand. What would be the fair 'entry fee' for this game? In other words, how much money would you be willing to pay in order to play this game? Perhaps you want to base this amount on the expectation of X. The idea would be that the expectation of X is the average amount of money that you recieve, and it would be only fair to pay exactly this amount in advance, making the game fair. However, we have

$$E(X) = \sum_{n=1}^{\infty} 2^n 2^{-n} = \infty,$$

and you would not be willing (or able) to pay an infinite amount of money to play this game. □

In Example 2.3.3, we computed the expectation of Y by first computing the probability mass function of Y. In this case, this was fairly easy, but in general this may not be so straightforward. Therefore, it would be nice if we can compute the expectation of a function of X by only using the probability mass function of X itself. This is indeed possible, and articulated in the following lemma.

Lemma 2.3.6. *Let X be a random variable, and $g : \mathbb{R} \to \mathbb{R}$. Then the expectation of $g(X)$ is given by*

$$E(g(X)) = \sum_{x} g(x) P(X = x),$$

whenever this last sum is well defined.

Proof. Since the sum is well defined, we are free to change the order of summation. We can now write

$$
\begin{aligned}
E(g(X)) &= \sum_{y} y P(g(X) = y) = \sum_{y} y \sum_{x:g(x)=y} P(X = x) \\
&= \sum_{y} \sum_{x:g(x)=y} y P(X = x) = \sum_{y} \sum_{x:g(x)=y} g(x) P(X = x) \\
&= \sum_{x} g(x) P(X = x).
\end{aligned}
$$

□

Example 2.3.7. For the random variable Y in Example 2.3.3, this result leads to the following computation: $E(Y) = \sum_{x} x^2 P(X = x) = 1 \cdot \frac{1}{3} + 1 \cdot \frac{1}{3} + 4 \cdot \frac{1}{3} = 2.$ □

♠ **Exercise 2.3.8.** Show that if X takes only non-negative integer values, we have

$$E(X) = \sum_{n=0}^{\infty} P(X > n).$$

For two random variables X and Y on the same sample space, we can define new random variables $X + Y$, XY, etcetera via

$$(X + Y)(\omega) = X(\omega) + Y(\omega), \text{ and } (XY)(\omega) = X(\omega)Y(\omega).$$

Expectations turn out to behave very nicely under taking sums of random variables. This will make it quite easy to compute the expectation of various random variables, as we shall see in the next section.

Theorem 2.3.9. *Let X and Y be two random variables defined on the same sample space. If $E(X)$ and $E(Y)$ are not both infinite with opposite signs, then*

$$E(X + Y) = E(X) + E(Y).$$

Proof. First,

$$
\begin{aligned}
E(Y) &= \sum_j j P(Y = j) = \sum_j j \sum_i P(X = i, Y = j) \\
&= \sum_i \sum_j j P(X = i, Y = j),
\end{aligned}
$$

and therefore,

$$
\begin{aligned}
E(X) + E(Y) &= \sum_i i P(X = i) + \sum_i \sum_j j P(X = i, Y = j) \\
&= \sum_i i \sum_j P(X = i, Y = j) + \sum_i \sum_j j P(X = i, Y = j) \\
&= \sum_i \sum_j (i + j) P(X = i, Y = j) \\
&= \sum_z \sum_j z P(X = z - j, Y = j) \\
&= \sum_z z \sum_j P(X + Y = z, Y = j) \\
&= \sum_z z P(X + Y = z).
\end{aligned}
$$

It follows that the last sum is well defined and hence $E(X + Y)$ exists and is equal to $E(X) + E(Y)$. $\qquad\square$

We make two remarks concerning the expectation of sums.

1. The conditions of the theorem are not necessary for the existence of $E(X + Y)$. For instance, when the random variable X satisfies $E(X) = \infty$, and $Y = -X$, then $E(X) + E(Y)$ is not defined, but $E(X + Y) = 0$.

2. One of the most common misconceptions in probability theory is the idea that Theorem 2.3.9 should only be true when X and Y are independent. However, the preceding calculation shows that the result has nothing to do with independence.

♠ **Exercise 2.3.10.** Extend Theorem 2.3.9 to more than two random variables.

Proposition 2.3.11. *For any random variable for which $E(X)$ exists and for any a and b, it is the case that $E(aX + b) = aE(X) + b$.*

♠ **Exercise 2.3.12.** Prove this proposition.

Instead of sums, we also need to consider products of random variables. It turns out that for products, independence does play a crucial role.

♠ **Exercise 2.3.13.** Find two random variables X and Y so that

$$E(XY) \neq E(X)E(Y).$$

Theorem 2.3.14. *If the random variables X and Y are independent and $E(X)$ and $E(Y)$ are finite, then $E(XY)$ is well defined and satisfies*

$$E(XY) = E(X)E(Y).$$

Proof. We write

$$
\begin{aligned}
\sum_l lP(XY = l) &= \sum_l l \sum_k P(X = k, Y = \frac{l}{k}) \\
&= \sum_k \sum_l lP(X = k, Y = \frac{l}{k}) \\
&= \sum_k \sum_l lP(X = k)P(Y = \frac{l}{k}) \\
&= \sum_{k \neq 0} kP(X = k) \sum_{l \neq 0} \frac{l}{k} P(Y = \frac{l}{k}) \\
&= E(X)E(Y).
\end{aligned}
$$

Hence the sum in the first line is well defined, and is therefore equal to $E(XY)$. □

It is *not* true that $E(XY) = E(X)E(Y)$ implies that X and Y are independent:

♠ **Exercise 2.3.15.** Let X and Y be independent random variables with the same distribution, taking values 0 and 1 with equal probability. Show that

$$E((X + Y)(|X - Y|)) = E(X + Y)E(|X - Y|),$$

but that $X + Y$ and $|X - Y|$ are not independent.

The expectation of a random variable X can be thought of as the average value taken by X. But clearly, this expectation does not say very much about the way the possible outcomes of X are spread out:

Example 2.3.16. Consider a random variable X with $P(X = 100) = 1/2$ and $P(X = -100) = 1/2$, and a random variable Y with $P(Y = 1) = 1/2$ and $P(Y = -1) = 1/2$. Clearly both $E(X)$ and $E(Y)$ are equal to zero, but it is also clear that outcomes of X will always be very far away from this expectation, while outcomes of Y will be much closer.   □

The *variance* of a random variable is intended to measure to what extent the outcomes of a random variable are spread-out. One way of doing this is to look at the 'average' deviation from the expectation, that is to the expectation of $|X - E(X)|$. It turns out that it is more convenient to look at the square $(X - E(X))^2$ instead.

Since it does not really make sense to talk about deviation from the expectation when this expectation is infinite, in the rest of this section, we assume that the expectations of the random variables involved are all finite.

Definition 2.3.17. Let X have finite expectation μ. The *variance* var(X) of X is defined as

$$\mathrm{var}(X) = E\left((X - \mu)^2\right) = \sum_x (x - \mu)^2 P(X = x).$$

The *standard deviation* $\sigma(X)$ of X is defined as the square root of the variance,

$$\sigma(X) = \sqrt{\mathrm{var}(X)}.$$

Lemma 2.3.18. $\mathrm{var}(X) = E(X^2) - (E(X))^2.$

Proof. We write $E(X) = \mu$. Then we have

$$
\begin{aligned}
\mathrm{var}(X) &= E(X^2 - 2\mu X + \mu^2) \\
&= E(X^2) - 2\mu E(X) + \mu^2 \\
&= E(X^2) - 2\mu^2 + \mu^2 = E(X^2) - \mu^2.
\end{aligned}
$$

□

Example 2.3.19. If X takes the values 1 and 0 with probability p and $1 - p$ respectively, then $E(X) = p$, $E(X^2) = p$ and therefore var$(X) = p - p^2$, using the last lemma.   □

Theorem 2.3.20. *Let X and Y be random variables, and $a, b \in \mathbb{R}$.*

(a) $\mathrm{var}(aX + b) = a^2\mathrm{var}(X).$

(b) $\mathrm{var}(X + Y) = \mathrm{var}(X) + \mathrm{var}(Y) + 2(E(XY) - E(X)E(Y))$. *In particular, if X and Y are independent, then*

$$\mathrm{var}(X + Y) = \mathrm{var}(X) + \mathrm{var}(Y).$$

Proof. (a) Let $E(X) = \mu$. We then write

$$
\begin{aligned}
\text{var}(aX + b) &= E((aX + b)^2) - (E(aX + b))^2 \\
&= E(a^2 X^2 + 2abX + b^2) - (a\mu + b)^2 \\
&= a^2 E(X^2) + 2ab\mu + b^2 - a^2 \mu^2 - 2ab\mu - b^2 \\
&= a^2 E(X^2) - a^2 \mu^2 = a^2 \text{var}(X).
\end{aligned}
$$

(b)

$$
\begin{aligned}
\text{var}(X + Y) &= E((X + Y - E(X + Y))^2) \\
&= E((X - E(X))^2 + (Y - E(Y))^2 + 2(XY - E(X)E(Y))) \\
&= \text{var}(X) + \text{var}(Y) + 2(E(XY) - E(X)E(Y)). \qquad \square
\end{aligned}
$$

Definition 2.3.21. The quantity $E(XY) - E(X)E(Y)$ which appears in (b) is called the *covariance* of X and Y, and denoted by $\text{cov}(X, Y)$.

♠ **Exercise 2.3.22.** Usually the covariance of X and Y is defined as $\text{cov}(X, Y) = E((X - E(X))(Y - E(Y)))$. Show that this amounts to the same definition as above.

♠ **Exercise 2.3.23.** Show that $\text{cov}(X, X) = \text{var}(X)$.

The expectation and variance of a random variable are frequently used objects. There exists a number of useful inequalities which are based on them.

Theorem 2.3.24. *Let X be a random variable taking only non-negative values. Then for any $a > 0$ we have*

$$
P(X \geq a) \leq \frac{1}{a} E(X).
$$

Proof.

$$
\begin{aligned}
E(X) &= \sum_x x P(X = x) \\
&\geq \sum_{x:x \geq a} x P(X = x) \geq a \sum_{x:x \geq a} P(X = x) \\
&= a P(X \geq a). \qquad \square
\end{aligned}
$$

♠ **Exercise 2.3.25.** Can you point out exactly where we have used the assumption that X takes only non-negative values?

Applying this result to $|X|^k$ leads to

Corollary 2.3.26 (Markov's inequality).

$$
P(|X| \geq a) \leq \frac{1}{a^k} E(|X|^k).
$$

Markov's inequality for $k = 2$ and applied to $|X - E(X)|$ leads to

Corollary 2.3.27 (Chebyshev's inequality).

$$P(|X - E(X)| \geq a) \leq \frac{1}{a^2}\mathrm{var}(X).$$

♠ **Exercise 2.3.28.** Show how the inequalities of Markov and Chebyshev follow from Theorem 2.3.24.

Finally, we mention the famous Cauchy–Schwarz inequality.

Theorem 2.3.29 (Cauchy–Schwarz inequality). *For any random variables X and Y for which $E(XY)$ is defined, we have*

$$E(XY) \leq \sqrt{E(X^2)E(Y^2)}.$$

Proof. The proof of this result is quite surprising. If $E(X^2) = \infty$ or $E(Y^2) = \infty$, then there is nothing to prove. If they are both finite, then we first claim that $E(XY)$ is well defined. To see this, observe that $|xy| \leq \frac{1}{2}(x^2 + y^2)$, for all real x and y.

♠ **Exercise 2.3.30.** Show that this last observation leads to the claim.

Also, if $E(X^2)$ or $E(Y^2)$ is zero, then X or Y takes the value zero with probability 1, and there is nothing to prove. Hence we can assume that the right hand side is positive. Now let a be a real number and define $Z = aX - Y$. Then

$$0 \leq E(Z^2) = a^2 E(X^2) - 2aE(XY) + E(Y^2).$$

The right hand side can be seen as a quadratic equation in the variable a. Since this quadratic expression is apparently non-negative, it follows that the corresponding discriminant is non-positive. That is, we have

$$(2E(XY))^2 - 4E(X^2)E(Y^2) \leq 0,$$

which is what we wanted to prove.   □

♠ **Exercise 2.3.31.** Show that we have equality in the Cauchy–Schwarz inequality if and only if $P(aX = Y) = 1$ for some a, or $X = 0$ with probability 1.

Here follows a number of examples of how to compute expectations and variances.

Example 2.3.32 (Binomial distribution). Recall that a random variable X is said to have a binomial distribution with parameters $n \in \mathbb{N}$ and $p \in [0, 1]$ if

$$P(X = k) = \binom{n}{k}p^k(1 - p)^{n-k},$$

for $k = 0, 1, \ldots, n$. The random variable X represents the number of heads when we flip a coin n times, where each flip gives heads with probability p. In particular, we can write such a random variable X as a sum

$$X = \sum_{i=1}^{n} Y_i,$$

where $Y_i = 1$ if the ith flip yields a head, and $Y_i = 0$ otherwise. It is clear that $E(Y_i) = p$, and according to the sum formula for expectations, we find that

$$E(X) = np. \qquad \square$$

♠ **Exercise 2.3.33.** Use the sum formula for variances to show that $\mathrm{var}(X) = np(1-p)$.

Here we see the convenience of the sum formulas. Computing the expectation of X directly from its probability mass function is possible but tedious work. The sum formula makes this much easier.

Example 2.3.34 (Poisson distribution). In Example 2.3.4 we showed that when X has a Poisson distribution with parameter λ, then $E(X) = \lambda$. We now compute its variance. For this we first compute $E(X^2)$. A nice trick makes things easier: instead of computing $E(X^2)$ directly, we compute $E(X(X-1)) = E(X^2) - E(X)$. The first equality in the next computation follows from Lemma 2.3.6.

$$
\begin{aligned}
E(X(X-1)) &= \sum_{k=0}^{\infty} k(k-1)e^{-\lambda}\frac{\lambda^k}{k!} \\
&= e^{-\lambda}\lambda^2 \sum_{k=2}^{\infty} \frac{\lambda^{k-2}}{(k-2)!} = \lambda^2.
\end{aligned}
$$

It follows that $E(X^2) = \lambda^2 + \lambda$ and hence, $\mathrm{var}(X) = E(X^2) - (E(X))^2 = \lambda$. $\quad\square$

Example 2.3.35 (Geometric distribution). Recall that a random variable X has a geometric distribution with parameter $p \in (0, 1]$ if

$$P(X = k) = p(1-p)^{k-1},$$

for $k = 1, 2, \ldots$ To compute its expectation, we write

$$E(X) = p\sum_{k=1}^{\infty} k(1-p)^{k-1}.$$

Let us denote $\sum_{k=1}^{n} k(1-p)^{k-1}$ by S_n, and $\sum_{k=1}^{\infty} k(1-p)^{k-1}$ by S. We would like to compute S, and there are various ways of doing this. We use an interesting trick, but see Exercise 2.3.36 for an alternative approach. The point is that we recognise S as the derivative (with respect to p) of $-\sum_{k=0}^{\infty}(1-p)^k$, and this last expression is equal to $-1/p$. It follows that $S = p^{-2}$. Note that we have used the fact that we are allowed to differentiate the series term by term, in fact we will state this formally in the forthcoming Theorem 2.6.6(b).   □

♠ **Exercise 2.3.36.** Here is an alternative way of computing S, without using differentiation. We can write $(1 - p)S_n$ as

$$(1 - p)S_n = (1 - p) + 2(1 - p)^2 + 3(1 - p)^3 + \cdots + n(1 - p)^n.$$

Hence,

$$pS_n = S_n - (1 - p)S_n = 1 + (1 - p) + (1 - p)^2 + \cdots + (1 - p)^{n-1} - n(1 - p)^n, \quad (2.1)$$

and we find that

$$S_n = \frac{1 + (1 - p) + (1 - p)^2 + \cdots + (1 - p)^{n-1}}{p} - \frac{n(1 - p)^n}{p}.$$

Now finish the argument yourself, by taking the limit for $n \to \infty$.

Example 2.3.37 (A distribution without an expectation). Let X have probability mass function

$$P(X = k) = \frac{C}{k^2},$$

for all $k \in \mathbb{Z}, k \neq 0$, and where C is chosen such the sum equals 1. This random variable does not have an expectation, since the sum

$$\sum_{k=-\infty, k\neq 0}^{+\infty} \frac{Ck}{k^2}$$

is not well defined.   □

2.4   Random Vectors

Apart from the discussion of independence, we have so far studies individual random variables. Very often however, the interplay between different random variables (defined on the same sample space) is very important, and not only in situations with independence.

Example 2.4.1. Suppose that I ask you to throw a coin twice. The outcome of this experiment is a vector, which we may denote by (X, Y), where X represents the first throw, and Y the second. Now suppose that there are two types of students: diligent and lazy. The diligent student really throws the coin twice, and his or her outcome can be any of the four possibilities (writing 0 for head and 1 for tail) $(0, 0), (0, 1), (1, 0)$ and $(1, 1)$. Clearly, we want the four outcomes to be equally likely. The lazy student however, throws only once and copies the outcome for the second coordinate of his outcome. His (or her) possible outcomes are $(0,0)$ and $(1,1)$, again with the same probability. Suppose now that we only look at the first coordinate X. In *both* cases, the probability that $X = 0$ is equal to $1/2$. Also, when we look at the second coordinate, the probability that $Y = 0$ is the same for the diligent and lazy student. Hence, by looking at the *individual* outcomes X and Y, we can not distinguish between the two types of student. The only way to distinguish the two students, is to look at the outcome of the *complete* vector (X, Y). $\square$

Definition 2.4.2. A *random vector* $(X_1, \ldots, X_d)$ is a mapping from a sample space Ω into $\mathbb{R}^d$.

Definition 2.4.3. The *joint probability mass function* of a random vector $X = (X_1, X_2, \ldots, X_d)$ is defined as

$$p_X(x_1, x_2, \ldots, x_d) = P(X_1 = x_1, \ldots, X_d = x_d).$$

The distribution of any of the individual X_i's is referred to as a *marginal* distribution, or just a *marginal*.

Definition 2.4.4. The *joint distribution function* of $X = (X_1, \ldots, X_d)$ is the function $F_X : \mathbb{R}^d \to [0, 1]$ given by

$$F_X(x_1, \ldots, x_d) = P(X_1 \leq x_1, \ldots, X_d \leq x_d).$$

In Example 2.4.1 it became clear that it is possible to have two random vectors (X, Y) and (V, W) so that X and V have the same marginal distribution, Y and W also have the same marginal distribution, but nevertheless the joint distributions are different. Hence we cannot in general find the joint distributions if we only know the marginals.

The next result shows that the opposite direction is possible: if we know the joint distribution, then we also know the marginal distributions. This reinforces the idea that the joint distribution really says more than the collection of marginal distributions.

Theorem 2.4.5. *Let $(X_1, X_2, \ldots, X_d)$ have probability mass function $p(x_1, \ldots, x_d)$. Then the mass function of X_1 can be written as*

$$p_{X_1}(x_1) = \sum_{x_2, x_3, \ldots, x_d} p(x_1, x_2, \ldots, x_d),$$

and similarly for the other marginals. In words, we find the mass function of X_1 by summing over all the other *variables.*

Proof. This is an immediate consequence of Theorem 1.4.11, where we take A to be the event that $X_1 = x_1$ and the B_i's all possible outcomes of the remaining coordinates. $\square$

♠ **Exercise 2.4.6.** Provide the details of the last proof.

Example 2.4.7. Let (X, Y) have joint probability mass function p given by $p(0, 0) = 0.4$, $p(0, 1) = 0.2$, $p(1, 0) = 0.1$ and $p(1, 1) = 0.3$. Then

$$P(X = 0) = p(0, 0) + p(0, 1) = 0.6$$

and

$$P(X = 1) = p(1, 0) + p(1, 1) = 0.4.$$

We can read off all information about the vector (X, Y) from the following table. The last column contains the marginal probability mass function of X, obtained by summing all probabilites row-wise. The last row contains the probability mass function of Y, obtained by summing the probabilities column-wise.

	0	1	X
0	0.4	0.2	0.6
1	0.1	0.3	0.4
Y	0.5	0.5	1

□

It turns out that the joint distribution comes in handy when deciding whether or not two or more random variables are independent. Indeed, the very definition of independence implies that X and Y are independent if and only if

$$p_{(X,Y)}(x, y) = p_X(x) p_Y(y),$$

for all x and y. The following result makes it sometimes even easier to decide whether or not X and Y are independent. The point is that the functions g and h below need not be the marginal mass functions of X and Y respectively.

Theorem 2.4.8. *The random variables X and Y are independent if and only if $p_{(X,Y)}$ can be factorised as a function of x and a function of y, that is, can be written as $p_{(X,Y)}(x, y) = g(x)h(y)$ for all (x, y).*

Proof. If X and Y are independent, then clearly we can take g and h as the marginal mass functions of X and Y. To prove the converse, we write p for the mass function of (X, Y). Suppose that

$$p(x, y) = g(x)h(y)$$

for all x and y. For any x, summing over the second coordinate and using Theorem 2.4.5 leads to

$$p_X(x) = \sum_{y_1} p(x, y_1) = g(x) \sum_{y_1} h(y_1).$$

Since $\sum_{x_1} p_X(x_1) = 1$, this leads to $\sum_{x_1} g(x_1) \sum_{y_1} h(y_1) = 1$. Similarly, we find for any y that

$$p_Y(y) = h(y) \sum_{x_1} g(x_1).$$

Hence, for given x and y we obtain

$$\begin{aligned} p_X(x)p_Y(y) &= g(x)h(y) \sum_{y_1} h(y_1) \sum_{x_1} g(x_1) \\ &= g(x)h(y) = p(x,y), \end{aligned}$$

which is what we wanted to prove. $\qquad\square$

Note that it follows from this proof that if $p(x,y) = g(x)h(y)$, then $g(x)$ differs from $p_X(x)$ by only a multiplicative constant, and similarly for $h(y)$ and $p_Y(y)$. So although $g(x)$ is not necessarily $p_X(x)$, the difference is only a matter of normalisation, and similarly for $h(y)$ and $p_Y(y)$.

Example 2.4.9. Let (X,Y) have joint mass function

$$p(k,l) = \frac{\lambda^k \mu^l}{k!l!} e^{-(\lambda+\mu)},$$

for $k, l = 0, 1, 2, \ldots$, and where $\lambda, \mu > 0$. It is obvious that $p(k,l)$ factorises as the product of a function of k and a function of l, and therefore X and Y are independent. To compute the marginal distribution of X we compute

$$\begin{aligned} P(X = k) &= \sum_{l=0}^{\infty} p(k,l) \\ &= e^{-(\lambda+\mu)} \frac{\lambda^k}{k!} \sum_{l=0}^{\infty} \frac{\mu^l}{l!} \\ &= e^{-(\lambda+\mu)} \frac{\lambda^k}{k!} e^{\mu} \\ &= e^{-\lambda} \frac{\lambda^k}{k!}, \end{aligned}$$

which we recognise as the probability mass function of a Poisson distribution with parameter λ. Similarly, Y has a Poisson distribution with parameter μ. $\qquad\square$

Example 2.4.10. Let (X,Y) have joint mass function

$$p(k,n) = \frac{C \cdot 2^{-k}}{n}, \quad \text{for } k = 1, 2, \ldots \text{ and } n = 1, \ldots, k,$$

and suitable constant C. It seems that $p(k,n)$ can be factorised as $p(k,n) = C2^{-k}\frac{1}{n}$ which would imply that X and Y are independent. However, this is not true. There is dependency between X and Y, which you can see when you look at the values taken by k and n. In fact, to avoid the range conditions, we could write

$$p(k,n) = \frac{C2^{-k}}{n} \cdot \mathbf{1}_{\{n \leq k\}}(k,n),$$

. where $\mathbf{1}_{\{n \leq k\}}(k, n)$ is an *indicator function* taking the value 1 if $n \leq k$ and 0 otherwise. Now it is clear that $p(k, n)$ cannot be factorised, and we see that X and Y are not independent. Of course, this conclusion also follows from the fact that $P(X = 1) > 0$, $P(Y = 2) > 0$ and $P(X = 1, Y = 2) = 0$. $\square$

We can also use joint mass functions to determine the distribution of a *sum* of two random variables.

Theorem 2.4.11. *Let (X, Y) be a random vector on some sample space, with mass function p, and let $Z = X + Y$. Then the mass function p_Z of Z is given by*

$$p_Z(z) = \sum_x p(x, z - x) = \sum_y p(z - y, y).$$

In particular, when X and Y are independent we find

$$p_Z(z) = \sum_x p_X(x) p_Y(z - x) = \sum_y p_X(z - y) p_Y(y).$$

♠ **Exercise 2.4.12.** Prove this theorem.

Example 2.4.13. Suppose X and Y are independent with joint mass function as in Example 2.4.9. This means, as noted before, that X and Y have a Poisson distribution with parameters λ and μ respectively. How can we compute the distribution of the sum $X + Y$? According to Theorem 2.4.11 and Proposition 1.5.9, we find

$$
\begin{aligned}
P(X + Y = z) &= \sum_{k=0}^{z} \frac{\lambda^k \mu^{z-k}}{k!(z-k)!} e^{-(\lambda+\mu)} \\
&= \frac{e^{-(\lambda+\mu)}}{z!} \sum_{k=0}^{z} \binom{z}{k} \lambda^k \mu^{z-k} \\
&= e^{-(\lambda+\mu)} \frac{(\lambda + \mu)^z}{z!}.
\end{aligned}
$$

This means that $X + Y$ again has a Poisson distribution, whose parameter is the sum of the original parameters. $\square$

We end this section with the analogue of Lemma 2.3.6 for vectors.

Lemma 2.4.14. *Let $(X_1, X_2, \ldots, X_d)$ be a random vector and $g : \mathbb{R}^d \to \mathbb{R}$. Then*

$$E(g(X_1, \ldots, X_d)) = \sum_{x_1, \ldots, x_d} g(x_1, \ldots, x_d) P(X_1 = x_1, \ldots, X_d = x_d),$$

whenever this sum is well defined.

♠ **Exercise 2.4.15.** Prove this lemma.

2.5 Conditional Distributions and Expectations

One of the most important concepts in probability theory is that of conditioning. Very often, we want to know something about a certain random variable when some information about another is available. In principle, we have dealt already with conditional probability in Chapter 1, but now we will talk about this in terms of random variables. In the following, X and Y are random variables on the same sample space.

Definition 2.5.1. The *conditional probability mass function* of Y given $X = x$ is defined as

$$p_{Y|X}(y|x) = P(Y = y|X = x),$$

whenever $P(X = x) > 0$. The *conditional distribution function* of a random variable Y given $X = x$ is defined as

$$F_{Y|X}(y|x) = P(Y \leq y|X = x),$$

whenever $P(X = x) > 0$.

It follows from this definition that

$$p_{Y|X}(y|x) = \frac{p_{(X,Y)}(x, y)}{p_X(x)}. \tag{2.2}$$

♠ **Exercise 2.5.2.** Derive formula (2.2).

When we learn that $X = x$, our conditional probability mass function for Y is $p_{Y|X}(y|x)$, which should be seen as a function of y. The corresponding expectation is then given by $\sum_y y p_{Y|X}(y|x)$. Notice that this conditional expectation is a function of x, the outcome of X. We denote this function of x by $E(Y|X = x)$.

Definition 2.5.3. The *conditional expectation* of Y given X is denoted by $E(Y|X = x)$, and defined as

$$E(Y|X = x) = \sum_y y p_{Y|X}(y|x).$$

Example 2.5.4. Consider the vector (X, Y) in Example 2.4.7. We have

$$\begin{aligned} p_{Y|X}(0|0) &= P(Y = 0|X = 0) \\ &= \frac{p(0, 0)}{P(X = 0)} = \frac{0.4}{0.6} = \frac{2}{3}. \end{aligned}$$

Similarly, $p_{Y|X}(1|0) = \frac{1}{3}$, and this leads to

$$E(Y|X = 0) = 0 \cdot \frac{2}{3} + 1 \cdot \frac{1}{3} = \frac{1}{3}. \qquad \square$$

♠ **Exercise 2.5.5.** Compute $E(Y|X = 1)$, $E(X|Y = 0)$ and $E(X|Y = 1)$ in this example.

Example 2.5.6. Consider, as in Example 2.4.13, two independent random variables X and Y, with Poisson distributions with parameters λ and μ respectively. We are interested in the conditional distribution of X, given $X + Y$. That is, we want to compute $P(X = k|X + Y = m + k)$. Keeping in mind that $X + Y$ has again a Poisson distribution with parameter $\lambda + \mu$, we find

$$
\begin{aligned}
P(X = k|X + Y = m + k) &= \frac{P(X = k, Y = m)}{P(X + Y = m + k)} \\
&= \frac{\frac{\lambda^k}{k!}e^{-\lambda}\frac{\mu^m}{m!}e^{-\mu}}{\frac{(\lambda+\mu)^{m+k}}{(m+k)!}e^{-(\lambda+\mu)}} \\
&= \frac{(m + k)!}{k!m!}\frac{\lambda^k\mu^m}{(\lambda + \mu)^{m+k}} \\
&= \binom{m + k}{k}\left(\frac{\lambda}{\lambda+\mu}\right)^k\left(1 - \frac{\lambda}{\lambda+\mu}\right)^m.
\end{aligned}
$$

This is to say that (take $m + k = r$)

$$
p_{X|X+Y}(k|r) = \binom{r}{k}\left(\frac{\lambda}{\lambda+\mu}\right)^k\left(1 - \frac{\lambda}{\lambda+\mu}\right)^{r-k},
$$

so that the conditional distribution of X given $X + Y = r$ is a binomial distribution with parameters r and $\lambda/(\lambda + \mu)$. Hence

$$
E(X|X + Y = r) = \frac{r\lambda}{\lambda + \mu}. \qquad \square
$$

Conditional expectations are extremely useful for computing unconditional expectations. This is articulated in the following theorem.

Theorem 2.5.7. *For random variables X and Y defined on the same sample space, we have*

$$
E(Y) = \sum_x E(Y|X = x)P(X = x).
$$

Proof.

$$
\begin{aligned}
E(Y) &= \sum_y y P(Y = y) \\
&= \sum_y y \sum_x P(X = x, Y = y) \\
&= \sum_y y \sum_x p_{Y|X}(y|x) p_X(x) \\
&= \sum_x p_X(x) \sum_y y p_{Y|X}(y|x) \\
&= \sum_x P(X = x) E(Y|X = x).
\end{aligned}
$$

$\square$

Example 2.5.8. A chicken produces N eggs, where N is a random variable with a Poisson distribution with parameter λ. Each of the eggs hatches with probability p, independently of the other eggs. Let K be the number of chicks. We want to find $E(K|N = n)$ and $E(K)$. To do this, note that the assumptions tell us that

$$
p_N(n) = \frac{\lambda^n}{n!} e^{-\lambda}
$$

and

$$
p_{K|N}(k|n) = \binom{n}{k} p^k (1 - p)^{n-k}.
$$

This gives us that $E(K|N = n) = \sum_k k p_{K|N}(k|n) = pn$, and hence

$$
\begin{aligned}
E(K) &= \sum_n E(K|N = n) P(N = n) \\
&= \sum_n pn P(N = n) = p \sum_n n P(N = n) \\
&= p E(N) = p\lambda.
\end{aligned}
$$

Note that we have computed $E(K)$ without computing the probability mass function of K itself. In fact, it is also possible to compute this probability mass function:

$$
\begin{aligned}
P(K = k) &= \sum_{n=k}^{\infty} P(K = k|N = n) P(N = n) \\
&= e^{-\lambda} \sum_{n=k}^{\infty} \frac{\lambda^n}{n!} \binom{n}{k} p^k (1 - p)^{n-k} \\
&= e^{-\lambda} \frac{(\lambda p)^k}{k!} \sum_{n=k}^{\infty} \frac{\lambda^{n-k}}{(n-k)!} (1 - p)^{n-k} \\
&= e^{-\lambda} \frac{(\lambda p)^k}{k!} e^{\lambda(1-p)} = e^{-\lambda p} \frac{(\lambda p)^k}{k!},
\end{aligned}
$$

which means that K has a Poisson distribution with parameter λp.

It is a little more work to compute $E(N|K = k)$. To do this, we assume that $n \geq k$ (why?) and write

$$
\begin{aligned}
P(N = n|K = k) &= \frac{P(N = n, K = k)}{P(K = k)} \\
&= \frac{P(K = k|N = n)P(N = n)}{P(K = k)} \\
&= \frac{e^{-\lambda}\frac{\lambda^n}{n!}\binom{n}{k}p^k(1-p)^{n-k}}{e^{-\lambda p}\frac{(\lambda p)^k}{k!}} \\
&= \frac{(\lambda(1-p))^{n-k}}{(n-k)!}e^{-\lambda(1-p)}.
\end{aligned}
$$

This we also recognise as the probability mass function of a Poisson random variable, and we now see that

$$
\begin{aligned}
E(N|K = k) &= \sum_{n=k}^{\infty} nP(N = n|K = k) \\
&= \sum_{n=k}^{\infty} (k + n - k)\frac{(\lambda(1-p))^{n-k}}{(n-k)!}e^{-\lambda(1-p)} \\
&= k + \lambda(1-p),
\end{aligned}
$$

where we use the fact that a random variable with a Poisson distribution with parameter $\lambda(1-p)$ has expectation $\lambda(1-p)$. □

♠ **Exercise 2.5.9.** Explain why this last result agrees with our probabilistic intuition. In retrospect, this is the answer which we could have anticipated.

The following amusing and important example shows that we must be extremely careful when we deal with conditional expectations and distributions.

Example 2.5.10 (First envelope problem). Suppose that I give you two envelopes, both containing a certain amount of money. I do not tell you the exact amounts, but I do tell you that one envelope contains twice the amount of the other. For convenience, we also assume that the amounts are powers of 2, that is, of the form 2^n for some $n \in \mathbb{Z}$. Now you have to choose an envelope, but you are not allowed to open it. After choosing this envelope I will give you the chance to swap. Does it make sense for you to do that? Clearly, common sense tells us that swapping couldn't make the slightest difference, but consider the following reasoning:

The amount of money in the chosen envelope is denoted by X, the amount in the other is denoted by Y. That is, we consider a random vector (X, Y). Now suppose that $X = x$. What is the conditional expectation of Y given this information? That is, we want to compute $E(Y|X = x)$. When $X = x$, Y can take only two values, namely $2x$ and $x/2$. Since you have chosen a random envelope, you have probability $1/2$ to have the envelope with the largest amount of money. That is, we have

$$P(Y = 2x | X = x) = P(Y = \frac{x}{2} | X = x) = \frac{1}{2}.$$

This leads to

$$E(Y | X = x) = \frac{1}{2} \cdot 2x + \frac{1}{2} \cdot \frac{x}{2} = \frac{5x}{4}.$$

This is very strange, since this conditional expectation of Y given $X = x$ is now strictly larger than x, the amount of your chosen envelope?! This suggests that you should swap, since the expected amount of the other envelope is larger than what you have in your hands.

There must be something wrong, and we will try to explain now what the trouble is. Consider the random vector (X, Y). The procedure of choosing the envelopes makes some implicit assumptions about the distribution of this random vector. In the first place, the marginal distributions of X and Y must be the same (why?). Also, as articulated above,

$$\begin{aligned} P(Y = x) &= P(Y = x | X = 2x)P(X = 2x) + P(Y = x | X = \frac{x}{2})P(X = \frac{x}{2}) \\ &= \frac{1}{2}P(X = 2x) + \frac{1}{2}P(X = \frac{x}{2}). \end{aligned}$$

Taking the last two observations together now gives

$$P(X = x) = \frac{1}{2}P(X = 2x) + \frac{1}{2}P(X = \frac{x}{2}).$$

Now recall that we have assumed that X takes only values of the form 2^n. Denote $P(X = 2^n)$ by q_n. The last display tells us that for all $n \in \mathbb{Z}$, we have

$$q_n = \frac{q_{n-1} + q_{n+1}}{2}.$$

A moment of thought reveals that this implies that all points (n, q_n), plotted in the plane, must lie on a straight line. If the slope of this straight line is not zero, then there will be q_n's with $q_n < 0$, which is clearly impossible, since q_n represents a probability. On the other hand, if the line has slope zero, then $q_n = c$ for some c. If $c = 0$, then $\sum_n q_n = 0$ which is impossible since they should sum up to 1, being the probability mass function of X. If $c > 0$ then $\sum_n q_n = \infty$, which is also not allowed, for the same reason. We conclude that a random vector with the properties implicitly assumed, does not exist.

This is a most remarkable conclusion. The point is that an experiment as described cannot be performed. Apparently, it is impossible to distribute amounts of money over the envelopes such that *no matter what I see in the first*, the amounts in the other are half or twice this amount, with equal probabilities. □

2.6 Generating Functions

In this section we have a special look at random variables that take values in $\mathbb{N}$. We shall introduce a new concept, the *generating function* of such a random variable. There are at least two good reasons for doing so.

In the first place, generating functions are a very convenient tool for all sorts of computations, that would be difficult and tedious without them. These computations have to do with sums of random variables, expectations and variances. For an application of this, see Section 6.5.

In the second place, they are a very good warm up for the concept of *characteristic functions* which are similar in nature but not restricted to integer-valued random variables. In this section, $X, Y, \ldots$ are random variables taking values in $\mathbb{N}$.

Definition 2.6.1. The *generating function* of X is defined as

$$
\begin{aligned}
G_X(s) &= E(s^X) \\
&= \sum_{n=0}^{\infty} p_X(n) s^n,
\end{aligned}
$$

for all $s \in \mathbb{R}$ for which this sum converges.

♠ **Exercise 2.6.2.** Show that $G_X(s)$ converges for at least all $s \in [0, 1]$.

Before we continue, here are a few examples.

Example 2.6.3. If $P(X = c) = 1$, then $G_X(s) = s^c$. □

Example 2.6.4 (Geometric distribution). If X has a geometric distribution with parameter p, then (recall again the geometric series)

$$
\begin{aligned}
G_X(s) &= \sum_{k=1}^{\infty} s^k p (1-p)^{k-1} \\
&= \frac{p}{1-p} \sum_{k=1}^{\infty} (s(1-p))^k \\
&= \frac{p}{1-p} \frac{s(1-p)}{(1 - s(1-p))} = \frac{sp}{1 - s + sp}.
\end{aligned}
$$
□

♠ **Exercise 2.6.5.** Show that the generating function of a random variable X with a Poisson distribution with parameter λ, is given by

$$
G_X(s) = e^{\lambda(s-1)}.
$$

Working with generating functions requires some knowledge of infinite power series. we do not aim to treat the general theory of such infinite series, and we will state all we need in one theorem, the proof of which can be found in most textbooks on analysis.

Theorem 2.6.6. *Let $G_a(s) = \sum_{n=0}^{\infty} a_n s^n$ be a power series, where $a = (a_0, a_1, a_2, \ldots)$ is a sequence of non-negative real numbers. Then:*

(a) *There exists a radius of convergence $R \geq 0$, such that the series converges for $|s| < R$, and diverges for $|s| > R$.*

(b) *We can differentiate $G_a(s)$ term by term for all $|s| < R$.*

(c) *If $G_b(s) = \sum_{n=0}^{\infty} b_n s^n$ is another power series with*

$$G_a(s) = G_b(s),$$

for all $|s| < R'$ for some $0 < R' < R$, then $a_n = b_n$ for all n.

(d) **(Abel's theorem)** *If $R \geq 1$, then*

$$\lim_{s \uparrow 1} G_a(s) = \sum_{n=0}^{\infty} a_n.$$

We will first deal with the calculation of expectation and variance. Since our random variables in this section are non-negative, we have no problem with the existence of the expectation. When we talk about the kth derivative at 1, $G^{(k)}(1)$, we mean $\lim_{s \uparrow 1} G^{(k)}(s)$.

Theorem 2.6.7. *Let X have generating function G. Then:*

(a) $E(X) = G'(1)$,

(b) $\mathrm{var}(X) = G''(1) + G'(1) - G'(1)^2$.

Proof. For (a) we take $s < 1$ and write, using Theorem 2.6.6(b) and (d),

$$G'(s) = \sum_{n=1}^{\infty} n s^{n-1} p_X(n)$$

$$\rightarrow \sum_{n=1}^{\infty} n p_X(n) = E(X),$$

as $s \uparrow 1$. For (b) we do something more general first:

♠ **Exercise 2.6.8.** Show, using the same idea as in the proof of (a), that for all $k = 1, 2, \ldots$

$$E(X(X-1) \cdots (X - k + 1)) = G^{(k)}(1).$$

The expression in (b) now follows, since

$$\begin{aligned}
\mathrm{var}(X) &= E(X^2) - (E(X))^2 \\
&= E(X(X-1)) + E(X) - (E(X))^2 \\
&= G''(1) + G'(1) - G'(1)^2.
\end{aligned}$$

$\square$

Example 2.6.9. Let X have a Poisson distribution with parameter λ. Then $G_X(s) = e^{\lambda(s-1)}$. Hence $G_X'(1) = \lambda$ and $G_X''(1) = \lambda^2$. It follows that $E(X) = \lambda$ and $\text{var}(X) = \lambda^2 + \lambda - \lambda^2 = \lambda$. $\qquad\square$

♠ **Exercise 2.6.10.** Find the expectation and variance of a geometric distribution using generating functions.

Generating functions can also be very helpful in studying sums of random variables.

Theorem 2.6.11. *If X and Y are independent, then*

$$G_{X+Y}(s) = G_X(s)G_Y(s).$$

Proof. Since X and Y are independent, so are s^X and s^Y, according to Theorem 2.2.8. Hence

$$
\begin{aligned}
G_{X+Y}(s) &= E(s^{X+Y}) = E(s^X s^Y) \\
&= E(s^X)E(s^Y) = G_X(s)G_Y(s).
\end{aligned}
$$

$\qquad\square$

♠ **Exercise 2.6.12.** Extend this result to any finite number of random variables.

Example 2.6.13. This result can be used to compute the generating function of the binomial distribution. Indeed, if X has a binomial distribution with parameters n and p, then X has the same distribution as the sum of n independent random variables $X_1, \ldots, X_n$, all with the same distribution, and $P(X_i = 1) = p = 1 - P(X_i = 0)$. Clearly, $G_{X_i}(s) = (1-p) + ps$, and hence,

$$G_X(s) = \prod_{i=1}^{n} G_{X_i}(s) = ((1-p) + ps)^n.\qquad\square$$

♠ **Exercise 2.6.14.** Use generating functions to prove that the sum of two independent random variables with a Poisson distribution, has again a Poisson distribution. Be very precise about what part of Theorem 2.6.6 you use in your argument.

2.7 Exercises

Exercise 2.7.1. Suppose that we throw a fair coin three times. Let X be the number of heads. Find the probability mass function and distribution function of X.

Exercise 2.7.2. Compute the expectation of X in the previous exercise.

Exercise 2.7.3. Compute $P(X > 2)$ when X has a Poisson distribution with parameter 3.

Exercise 2.7.4. Compute $P(X > 10)$ when X has a geometric distribution with parameter p.

Exercise 2.7.5. Compute $P(X = 3)$ when X has a negative binomial distribution with parameters 2 and p.

Exercise 2.7.6. An urn contains 8 white, 4 black, and two red balls. We win 2 euro for each black ball that we draw, and lose 1 euro for each white ball that we draw. We choose three balls from the urn, and X denote our winnings. Write down the probability mass function of X.

Exercise 2.7.7. Consider an urn with 5 red and 4 white balls. We draw a random ball, look at the colour, and put the ball back into the urn, together with an extra ball of the same colour. After this, we again draw a random ball from the urn. The number of different colours that we have seen is denoted by X.
(a) Compute $E(X)$.
(b) Compute the conditional probability that the first ball was red, given that the second ball was white.

Exercise 2.7.8. We throw 10 times with a fair twelve-sided die, on which the numbers $1, \ldots, 12$ are written. Let X be the largest outcome among the 10 throws. Let A be the event that all 10 throws give the same number.
(a) Compute $P(X \leq 7)$.
(b) Compute $P(X = 7)$.
(c) Are A and $\{X \leq 3\}$ independent events?
(d) Let Z be the outcome of the first throw. Are A and $\{Z = 3\}$ independent events?
(e) Let Y be the largest outcome among the first 9 throws. Compute $P(X = k|Y = 6)$ for all relevant values of k.

Exercise 2.7.9. Suppose we throw a die twice. Let X be the product of the two outcomes.
(a) Compute the probability mass function of X.
(b) Compute $P(X = k|D)$, where D is the event that the sum of the outcomes is equal to 7.

Exercise 2.7.10. Let X be a random variable with probability mass function given by $p(-4) = p(4) = \frac{1}{16}$ and $p(2) = p(-2) = \frac{7}{16}$. Let $Y = aX + b$, for certain a and b.
(a) For what values of a and b does Y take values in $\{0, 1, 2, 3, 4\}$?
(b) Compute the probability mass function of Y for these choices of a and b.

Exercise 2.7.11. The NBA (National Basketball Association) draft lottery involves the 11 teams that had the worst won-lost records during the year. A total of 66 balls are placed in an urn. Each of these balls is inscribed with the name of a team: 11 have the name of the team with the worst record, 10 have the name of the team with the second worst record, and so on. A ball is then chosen at random and the team whose name is on the ball is given the first pick in the draft of players about to enter the league. Another ball is then chosen, and if it belongs to a different team, then the team to which it belongs receives the second draft pick. (If the ball belongs to the team receiving the first pick, then this ball is discarded and another one is chosen, and so on, until another team is chosen.) Finally, another ball is chosen and the team named on the ball (provided it is different from the previous two teams) receives the third pick. The remaining draft picks 4 through 11 are then awarded to the 8 teams that did not win the lottery in inverse order of their won-lost record.

Let X denote the draft pick of the team with the worst record. Find the probability mass function of X.

Exercise 2.7.12. We throw a fair die twice. Let X be the sum of the outcomes, and let Y be the highest outcome that we see. Finally, A is the event that the outcome of the first throw is higher than the outcome of the second.
(a) Compute the probability mass function of Y and compute $E(Y)$.
(b) Are X and Y independent?
(c) Are $\{X = 5\}$ and A independent events?

Exercise 2.7.13. An urn contains n balls, numbered $1, \ldots, n$. We take (without replacement and without ordering) 5 balls from the urn (assuming that $n \geq 5$). We let X denote the lowest number we have drawn, and Y the one but largest.
(a) Compute the probability mass functions of X and Y.
(b) Show that

$$\sum_{k=m}^{n} \binom{k}{m} = \binom{n+1}{m+1},$$

for all $n = 0, 1, \ldots$ and $m \leq n$.
(c) Use (b) to compute $E(X)$. It is convenient to use Exercise 2.3.8.

Exercise 2.7.14. Let X be a random variable with probability mass function

$$p_X(n) = c\frac{1}{n!},$$

for $n = 0, 1, 2, \ldots$, and $p_X(n) = 0$ otherwise.
(a) Compute c. (Use the expansion $e^x = \sum_{k=0}^{\infty} x^k/k!$)
(b) Compute the probability that X is even.
(c) Compute the expectation of X.

Exercise 2.7.15. Suppose we want to test a large number of blood samples in order to see if they contain a certain antibody. To reduce the amount of work, one proceeds as follows. We divide the samples into groups of size k, and these k samples are put together. The resulting mixtures are tested. If the test of such a mixture is negative, no further action is required. If it is positive, then the k original samples are individually tested after all, so that in such case, a total of $k+1$ tests needs to be performed. The samples contain the antibody with probability p, independently of each other.
(a) What is the probability that a mixture of k samples contains the antibody?
(b) Let S be the total number of tests that needs to be performed when the original

number of samples is $n = mk$. Compute $E(S)$ and $\text{var}(S)$.
(c) For what values of p does this method give an improvement, for suitable k when we compare this to individual tests right from the beginning? Find the optimal value of k as a function of p.

Exercise 2.7.16. Suppose that X has a geometrical distribution with parameter p. Show that
$$P(X = m + k | X > m) = P(X = k).$$
Explain why this property is called the *lack of memory property* of the geometrical distribution.

Exercise 2.7.17. Compute the expection of X when X has a negative binomial distribution.

Exercise 2.7.18. A gambling book recommends the following strategy for the game of roulette. It recommends that a gambler bets 1 euro on red. If red appears (with probability $\frac{18}{38}$), then the gambler should take his or her profit and quit the game. If the gambler loses his or her bet (which has probability $\frac{20}{38}$ of occurring), he should make additional 1 euro bets on red on each of the next two spins of the roulette wheel and then quit. Let X denote the gambler's winnings when he quits.
(a) Find $P(X > 0)$.
(b) Is this a winning strategy? Explain your answer.
(c) Compute $E(X)$.

Exercise 2.7.19. Suppose that X has a Poisson distribution with parameter λ. Show that
$$P(X = k) = \frac{\lambda}{k} P(X = k - 1),$$
for $k = 1, 2, \ldots$ Use this to find k so that $P(X = k)$ is maximal.

Exercise 2.7.20. Suppose that the number of children N of a randomly chosen family satisfies
$$P(N = n) = \frac{3}{5} \left(\frac{2}{5} \right)^n,$$
for $n = 0, 1, \ldots$.
(a) What is the probability that a family has no children?
(b) Compute the expectation of N.
Now suppose that a child is equally likely to be a girl or a boy, and let X be the number of daughters in a randomly chosen family.
(c) Compute $E(X|N = n)$ and use this to compute $E(X)$.

Exercise 2.7.21. Let (X, Y) be a random vector with probability mass function $p_{(X,Y)}(i, j) = 1/10$, for $1 \leq i \leq j \leq 4$.
(a) Show that this is indeed a probability mass function.
(b) Compute the marginal distributions of X and Y.
(c) Are X and Y independent?
(d) Compute $E(XY)$.

Exercise 2.7.22. Compute $E(X|Y = y)$ and $E(Y|X = x)$ in the previous exercise.

Exercise 2.7.23. Suppose that 15 percent of the families in a certain country have no children, 20 percent have 1, 35 percent have 2, and 30 percent have 3. Suppose further that each child is equally likely to be a boy or a girl, independent of the other children. A family is chosen at random, and we write B for the number of boys in this family, and G for the number of girls. Write down the joint probability mass function of (B, G).

Exercise 2.7.24. We roll two fair dice. Find the joint probability mass function of X and Y when
(a) X is the largest value obtained, and Y is the sum of the values;
(b) X is the value on the first die, Y is the largest value;
(c) X is the smallest value, and Y is the largest.

Exercise 2.7.25. Compute $E(Y|X = x)$ for all random variables in the previous exercise.

Exercise 2.7.26. Compute $E(X|Y = y)$ in Example 2.4.10.

Exercise 2.7.27. Compute $E(Y|X = x)$ for the lazy and the diligent student in Example 2.4.1.

Exercise 2.7.28. Suppose that we choose three different numbers randomly from $\{1, 2, \ldots, 10\}$. Let X be the largest of these three numbers.
(a) Compute the distribution function of X.
(b) Compute $P(X = 9)$.

Exercise 2.7.29. Suppose that a given experiment has k possible outcomes, the ith outcome having probability p_i. Denote the number of occurrences of the ith outcome in n independent experiments by N_i. Show that

$$P(N_i = n_i, i = 1, \ldots, k) = \frac{n!}{n_1! n_2! \cdots n_k!} p_1^{n_1} p_2^{n_2} \cdots p_k^{n_k}.$$

This is called the *multinomial distribution*.

Exercise 2.7.30. Compute the marginal distributions of the multinomial distribution.

Exercise 2.7.31. Suppose that X and Y are independent, and both have the *uniform distribution* on $\{0, 1, , \ldots, n\}$, that is, $P(X = i) = P(Y = i) = 1/(n+1)$, for all $i = 0, \ldots, n$. Show that

$$P(X + Y = k) = \frac{n + 1 - |n - k|}{(n+1)^2},$$

for all $k = 0, 1, \ldots, 2n$.

Exercise 2.7.32. Let X and Y be independent and geometrically distributed with the same parameter p. Compute the probability mass function of $X - Y$. Can you also compute $P(X = Y)$ now?

Exercise 2.7.33. Let X and Y be as in the previous exercise. Compute $E(X|X + Y = k)$ for all $k = 2, 3, \ldots$

Exercise 2.7.34. Let X be a random variable taking values in $\mathbb{N}$. Assume that $P(X = k) \geq P(X = k + 1)$, for all k. Show that

$$P(X = k) \leq \frac{2E(X)}{k^2}.$$

Exercise 2.7.35. We throw a die repeatedly, until the total sum of the outcomes is at least 1400. Let β be the probability that we need more than 420 throws of the die for this. Use Chebyshev's inequality to obtain a bound for β.

Exercise 2.7.36. Let $X_1, X_2, \ldots, X_n$ be independent random variables with the same distribution, with finite variance σ^2 and *negative* expectation μ. Let $S_n = X_1 + \cdots + X_n$ be their sum. Use Chebyshev's inequality to show that for any constant c, we have

$$\lim_{n \to \infty} P(S_n \geq c) = 0.$$

Exercise 2.7.37. Let X and Y be independent binomial random variables with parameters n and p. Denote their sum by Z. Show that

$$P(X = k|Z = m) = \frac{\binom{n}{k}\binom{n}{m-k}}{\binom{2n}{m}},$$

for $k = 1, \ldots, m$.

Exercise 2.7.38. Let $X_1, X_2, \ldots, X_n$ be independent random variables with $P(X_i = 1) = p_i$ and $P(X_i = 0) = 1 - p_i$, for all $i = 1, \ldots n$. We write $Y = X_1 + \cdots + X_n$.
(a) Show that $E(Y) = \sum_{i=1}^n p_i$.
(b) Show that $\text{var}(Y) = \sum_{i=1}^n p_i(1 - p_i)$.
(c) Show that, for $E(Y)$ fixed and equal to $E(Y) = 1$, the variance $\text{var}(Y)$ is maximal when all the p_i's are the same: $p_1 = \cdots = p_n = 1/n$.
(d) The result in (c) seems counterintuitive: the variance of the sum is greatest if the individuals are most alike. Try to explain this phenomenon.

Exercise 2.7.39 (Secretary's problem). A secretary drops n matching pairs of letters and envelopes down the stairs, and then places the letters into the envelopes in a random order. Let X be the number of correctly matched pairs.
(a) Show that the probability that a given envelope contains the correct letter is equal to $1/n$.
(b) Use (a) to show that the expectation of X is equal to 1.

(c) Show that the variance of X is also equal to 1.
(d) Show that for all k,

$$P(X = k) \rightarrow \frac{e^{-1}}{k!},$$

as $n \rightarrow \infty$.

Exercise 2.7.40. Suppose that X and Y are random variables on the same sample space, taking values in $\mathbb{N}$. Suppose that $X(\omega) \leq Y(\omega)$, for all $\omega \in \Omega$. Show that $E(X) \leq E(Y)$.

Exercise 2.7.41. Suppose we throw a coin 5 times. Let X be the number of heads, and Y be the number of tails. We assume that the coin is biased, and that the probability of head is equal to $1/3$. We are interested in $Z = X - Y$.
(a) Express Z as a function of X.
(b) Compute $E(Z)$ without computing the probability mass function of Z.
(c) Compute the probability mass function of Z, and use this to compute $E(Z)$ once again.

Exercise 2.7.42. Suppose that the number of times that a person gets sick during one year, has a Poisson distribution with expectation 5. A new medicine reduces this expectation to 3 for 75% of the population, and has no effect on the remaining 25%. Suppose that John Smith uses the medicine, and gets sick twice. What is the conditional probability that the medicine has an effect on him?

Exercise 2.7.43. A die is thrown 5 times. Use generating functions to compute the probability that the sum of the scores is 15.

Exercise 2.7.44. Compute the generating function of the negative binomial distribution of Example 2.1.11. (Recall that this distribution is obtained as a sum of geometrical random variables, and use Example 2.6.4.) Use this to compute the expectation and the variance of a negative binomial distribution.

Chapter 3

Random Walk

In this chapter, we concentrate on an experiment that is strongly related to coin flips, the *random walk* . We shall prove a number of results that are very surprising and counterintuitive. It is quite nice that such results can be proved at this point already. The proofs are all based on (ingenious) counting methods.

3.1 Random Walk and Counting

Consider a particle on the one-dimensional line, starting at position $a \in \mathbb{Z}$, say. This particle performs n random steps as follows. At each time $t = 1, 2, \ldots, n$, we flip a coin. If heads comes up, the particle moves one unit to the right, if tails comes up, the particle moves one unit to the left. The sample space corresponding to this experiment is simply $\Omega = \{-1, +1\}^n$, the set of sequences of length n with the symbols -1 and $+1$. For instance, if $n = 4$, and $\omega = (1, 1, -1, 1)$, then the first and second step are to the right, the third is to the left, and the fourth is to the right again. The reason that we define Ω as $\{-1, 1\}^n$ rather than $\{0, 1\}^n$, is that in the former case, we can conveniently represent the position of the particle after k steps by the random variable $S_k : \Omega \to \mathbb{R}$ as

$$S_k(\omega) = a + \sum_{i=1}^{k} \omega_i,$$

for $k = 1, 2, \ldots, n$. The probability measure on Ω associated to the random walk gives equal probability 2^{-n} to all 2^n possible outcomes. Hence, if we want to compute the probability of an event A, we need to count the number of elements in A and multiply by 2^{-n}. We should denote this probability measure by P_n, but we will drop the subscript n when no confusion is possible. We will mostly be interested in the joint distribution of the S_k's, that is, in the random vector

$$S(\omega) = (S_1(\omega), \ldots, S_n(\omega)) \, .$$

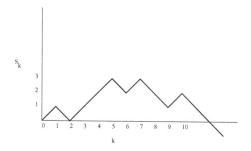

Figure 3.1: The graph of a random walk with $S_0 = 0, S_1 = 1, S_2 = 0$, etcetera.

A convenient way to think about this random vector, is to construct the graph obtained by joining the points $\{(k, S_k), k = 1, \ldots, n\}$, see Figure 3.1. We shall refer to such a graph as a *path*. Many important events can be described in terms of such paths.

The first interesting question deals with the marginal distribution of S_n. When the walk starts in a, then S_n cannot be equal to b if (i) n is odd and $b - a$ is even, or (ii) if n is even and $b - a$ is odd. Since we do not want to mention this all the time, we make the convention that $\binom{m}{r}$ is equal to 0 whenever r is not an integer. In the formulas below, this will then automatically imply that impossible events get probability 0 (check this for yourself!).

Theorem 3.1.1. *For all* $-n + a \leq b \leq n + a$, *we have*

$$P(S_n = b) = \binom{n}{\frac{1}{2}(n + b - a)} 2^{-n}.$$

Proof. We need to count the number of outcomes $\omega \in \Omega$ for which $S_n(\omega) = b$. For the sum of n ± 1s to be equal to $b - a$, it is necessary and sufficient to have $\frac{1}{2}(n + b - a)$ 1s, and $\frac{1}{2}(n - b + a)$ -1s. (To see how to obtain these numbers, denote the number of 1s by n_+ and the number of -1s by n_-. Then $n_+ + n_- = n$ and $n_+ - n_- = b - a$. Solving these equations leads to the answer given above.) There are $\binom{n}{\frac{1}{2}(n + b - a)}$ outcomes with this property, and the result follows immediately.   □

♠ **Exercise 3.1.2.** Give the marginal distribution of S_k for $k < n$.

♠ **Exercise 3.1.3.** Why is $P(S_n = b) = 0$ when $|b - a| > n$?

The following definition turns out to be very convenient for counting purposes.

Definition 3.1.4. We denote by $N_n(a, b)$ the number of paths that start in $(0, a)$ and end in (n, b), and by $N_n^0(a, b)$ the number of such paths which contain at least one point on the x-axis.

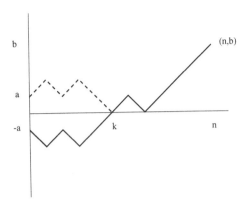

Figure 3.2: The reflection principle.

♠ **Exercise 3.1.5.** Show, using Theorem 3.1.1, that

$$N_n(a,b) = \binom{n}{\frac{1}{2}(n+b-a)}.$$

Lemma 3.1.6 (Reflection principle). *For $a, b > 0$ we have*

$$N_n^0(a,b) = N_n(-a,b).$$

Proof. Consider a path from $(0, -a)$ to (n, b). This path must intersect the x-axis for the first time at some time k, say. Now reflect the segment of the path between $x = 0$ and $x = k$ in the x-axis. The reflection is a path from $(0, a)$ to (n, b) which intersects the x-axis, see Figure 3.1. This reflection operator gives in fact a one-one correspondence between the appropriate collection of paths, and we are done. $\square$

♠ **Exercise 3.1.7.** Show that the reflection operator indeed gives a one-one correspondence.

Theorem 3.1.8 (Ballot theorem). *Let $b > 0$. The number of paths from $(0,0)$ to (n, b) which do not visit the x-axis (apart from the starting point) is equal to*

$$\frac{b}{n} N_n(0,b).$$

Proof. Note that the first step of such a path is necessarily to $(1, 1)$. After this first step, we need to make $n-1$ additional steps, from $(1, 1)$ to (n, b) in such a way that the path does not visit the x-axis. This can be done in $N_{n-1}(1, b) - N_{n-1}^0(1, b)$ ways. According to the reflection principle, this is equal to $N_{n-1}(1, b) - N_{n-1}(-1, b)$, which can be computed as follows.

$$
\begin{aligned}
N_{n-1}(1,b) - N_{n-1}(-1,b) &= \binom{n-1}{\frac{1}{2}(n-1+b-1)} - \binom{n-1}{\frac{1}{2}(n-1+b+1)} \\
&= \frac{(n-1)!}{(\frac{1}{2}(n+b-2))!(\frac{1}{2}(n-b-2))!} \times \\
&\quad \times \left\{ \frac{1}{\frac{1}{2}(n-b)} - \frac{1}{\frac{1}{2}(n+b)} \right\} \\
&= \frac{n!}{(\frac{1}{2}(n-b))!(\frac{1}{2}(n+b))!} \cdot \frac{b}{n} \\
&= \frac{b}{n} \binom{n}{\frac{1}{2}(n+b)}.
\end{aligned}
$$

$\square$

♠ **Exercise 3.1.9.** Interpret the special case $b = n$.

♠ **Exercise 3.1.10.** Do you understand why this is called the ballot theorem? Think of the following question: suppose that in a ballot, candidate A scores a votes, and candidate B scores b votes, where $a > b$. What is the probability that during the counting of the votes, A was always ahead of B?

Theorem 3.1.11. *Suppose that the random walk starts in 0, and let $b \neq 0$. Then we have*

$$
P(S_n = b, S_k \neq 0, \text{ for all } 1 \leq k < n) = \frac{|b|}{n} P(S_n = b).
$$

Proof. Take $b > 0$ (the case $b < 0$ is the subject of the next exercise.) The event at the left hand side occurs precisely when the path of the random walk does not visit the x-axis in the time interval from 1 to n, and $S_n = b$. The number of such paths is, according to the ballot theorem, equal to $\frac{b}{n} N_n(0,b)$. Now $N_n(0,b)$ is the number of paths that have $S_n = b$. Since all paths have the same probability, the result follows. $\square$

♠ **Exercise 3.1.12.** Give the proof of the last theorem for the case $b < 0$.

♠ **Exercise 3.1.13.** Show that

$$
P(S_1 S_2 \cdots S_n \neq 0) = \frac{1}{n} E(|S_n|).
$$

We end this section with a consequence of Theorem 3.1.11 which is interesting in its own right, but also very useful in the next section.

Lemma 3.1.14. *For the random walk we have, for all $m \geq 0$,*

$$
P(S_1 S_2 \cdots S_{2m} \neq 0) = P(S_{2m} = 0).
$$

Proof.

$$P(S_1 S_2 \cdots S_{2m} \neq 0) = \sum_{b=-2m}^{2m} P(S_1 S_2 \cdots S_{2m} \neq 0, S_{2m} = b)$$

$$= \sum_{b=-2m}^{2m} \frac{|b|}{2m} P(S_{2m} = b)$$

$$= 2 \sum_{k=1}^{m} \frac{2k}{2m} P(S_{2m} = 2k) = 2 \sum_{k=1}^{m} \frac{2k}{2m} \binom{2m}{m+k} 2^{-2m}$$

$$= 2 \cdot 2^{-2m} \sum_{k=1}^{m} \left\{ \binom{2m-1}{m+k-1} - \binom{2m-1}{m+k} \right\},$$

where we interpret $\binom{2m-1}{2m}$ as being equal to 0. The last expression contains a so called *telescoping sum*: when you write down all terms in the sum, you see that most terms cancel. The only terms that do not cancel are $\binom{2m-1}{m}$, coming from $k = 1$, and $-\binom{2m-1}{2m}$, coming from $k = m$. Since the latter is defined to be 0, we find that

$$P(S_1 S_2 \cdots S_{2m} \neq 0) = 2 \cdot 2^{-2m} \binom{2m-1}{m} = \binom{2m}{m} \left(\frac{1}{2}\right)^{2m}$$

$$= P(S_{2m} = 0). \qquad \square$$

3.2 The Arc-Sine Law

In this section, it is convenient to assume that the total number of steps of the random walk is even, that is, we assume that the total number of steps is equal to $2n$ for some $n \in \mathbb{N}$.

One might be inclined to think that if the particle makes $2n$ steps, starting in 0, the (random) *last* time at which the particle visits the x-axis tends to be closer to the end than to the beginning of the path. However, the following very surprising result shows that the distribution of this last visit to the x-axis, is in fact symmetric around n, the midpoint of the time interval. So, for instance, the probability that the last visit occurs at time 2 is the same as the probability that the last visit occurs at time $2n - 2$. I think that this is very counterintuitive.

The (random) last visit to 0, up to time $2n$, is denoted by L_{2n}.

Theorem 3.2.1. *For all $0 \leq k \leq n$ we have*

$$P(L_{2n} = 2k) = P(S_{2k} = 0)P(S_{2n-2k} = 0).$$

In particular, the distribution of L_{2n} is symmetric around n.

Proof. We have that $L_{2n} = 2k$ if and only if $S_{2k} = 0$ and there are no further visits to zero after that point, that is, if $S_{2k+1}S_{2k+2}\cdots S_{2n} \neq 0$. Hence,

$$
\begin{aligned}
P(L_{2n} = 2k) &= P(S_{2k} = 0, \, S_{2k+1}S_{2k+2}\cdots S_{2n} \neq 0)\\[2mm]
&= P\left(S_{2k} = 0, \, \sum_{i=2k+1}^{r} \omega_i \neq 0, \text{ for all } r = 2k+1,\dots,2n\right)\\[2mm]
&= P(S_{2k} = 0)P\left(\sum_{i=2k+1}^{r} \omega_i \neq 0, \text{ for all } r = 2k+1,\dots,2n\right)\\[2mm]
&= P(S_{2k} = 0)P\left(\sum_{i=1}^{r} \omega_i \neq 0, \text{ for all } r = 1,\dots,2n-2k\right)\\[2mm]
&= P(S_{2k} = 0)P(S_1 S_2 \cdots S_{2n-2k} \neq 0)\\[2mm]
&= P(S_{2k} = 0)P(S_{2n-2k} = 0),
\end{aligned}
$$

where the last equality follows from Lemma 3.1.14.   □

It is perhaps not so clear why this result is related to (or sometimes even called) the arc-sine law. I will explain this now. For this, we need the following classical asymptotic result, which I give without proof. A proof can be found in Feller (1978). For two sequences of real numbers a_n and b_n, $n = 1, 2, \dots$, both converging to infinity as $n \to \infty$, we say that $a_n \sim b_n$ if $a_n/b_n \to 1$ as $n \to \infty$.

Lemma 3.2.2 (Stirling's formula). *We have*

$$
n! \sim n^n e^{-n}(2\pi n)^{1/2}
$$

as $n \to \infty$. Moreover, it is the case that for all n,

$$
n^n e^{-n}(2\pi n)^{1/2} \leq n! \leq 2n^n e^{-n}(2\pi n)^{1/2}.
$$

The next theorem explains the title of this section. It says something about the behaviour of the random variable L_{2n} when n is large. Clearly, when n tends to infinity, one expects that the moment of the last visit to 0 also typically gets bigger and bigger. Therefore, in order to obtain an interesting result, we have to *normalise* L_{2n}. In this case, it turns out that we have to divide by $2n$.

Theorem 3.2.3 (Arc-sine law). *For $x \in (0,1)$, we have*

$$
\lim_{n \to \infty} P_{2n}\left(\frac{L_{2n}}{2n} \leq x\right) = \frac{2}{\pi}\arcsin x^{1/2}.
$$

This type of convergence will later in this book be called *convergence in distribution*. It is a notion of convergence in terms of distribution functions. In the proof of this theorem, we will need a technical result about Riemann integrals, which may not be so well known. The theorem provides a condition under which we can interchange limit and integral, and we give it without proof.

Theorem 3.2.4 (Dominated convergence). *Let $a < b$ and f, f_n, $n = 1, 2, \ldots$ be absolutely Riemann integrable functions with $f_n(x) \to f(x)$ as $n \to \infty$, for all $x \in (a, b)$. Suppose in addition that there exists a nonnegative function g with $\int_a^b g(x)dx < \infty$ and such that $|f_n(x)| \leq g(x)$, for all $x \in (a, b)$ and all n. Then*

$$\lim_{n \to \infty} \int_a^b f_n(x)dx = \int_a^b f(x)dx.$$

Proof of Theorem 3.2.3. Let $k \leq n$. Since

$$P_{2n}(S_{2k} = 0) = \binom{2k}{k} 2^{-2k} = \frac{(2k)!}{k!k!} 2^{-2k},$$

we can apply Stirling's formula to obtain that

$$P_{2n}(S_{2k} = 0) \sim \frac{1}{(\pi k)^{1/2}},$$

as $k \to \infty$ (and n as well, of course).

♠ **Exercise 3.2.5.** Prove this last statement.

It follows from this and Theorem 3.2.1, that

$$P_{2n}(L_{2n} = 2k) \sim \frac{1}{\pi(k(n-k))^{1/2}},$$

when both $k \to \infty$ and $n - k \to \infty$. Choose $x \in (0, 1)$ and choose a sequence of k, n such that $k/n \to x$. This implies that

$$\frac{n}{\pi(k(n-k))^{1/2}} \to \frac{1}{\pi(x(1-x))^{1/2}},$$

simply divide both the numerator and denominator on the left by n. Combining the last two displayed formulas now leads to

$$nP_{2n}(L_{2n} = 2k) \to \frac{1}{\pi(x(1-x))^{1/2}}, \tag{3.1}$$

as k, n tend to infinity in such a way that $k/n \to x$. Now consider the normalised random variable $L_{2n}/2n$, and define a function f_n on $(0, 1)$ in two steps as follows. First we define

$$f_n\left(\frac{k}{n}\right) = nP\left(\frac{L_{2n}}{2n} = \frac{k}{n}\right),$$

and then we define f_n between these points by

$$f_n(x) = f_n\left(\frac{k_x}{n}\right),$$

where $k_x = \max\{k; k/n \leq x\}$.

It follows from (3.1) that

$$f_n(x) \rightarrow \frac{1}{\pi(x(1-x))^{1/2}} =: f(x),$$

for all $x \in (0,1)$. According to the second statement in Lemma 3.2.2, on any interval $0 < u < x < v < 1$ there is a constant C so that

$$|f_n(x)| \leq \frac{C}{(x(1-x))^{1/2}}.$$

The right hand side is an integrable function and it then follows from Theorem 3.2.4 that

$$\int_u^v f_n(x)dx \rightarrow \int_u^v f(x)dx. \tag{3.2}$$

Now consider the distribution of $L_{2n}/2n$:

$$P\left(u < \frac{L_{2n}}{2n} < v\right) = \sum_{k:u<k/n<v} P\left(\frac{L_{2n}}{2n} = \frac{k}{n}\right)$$

$$= \frac{1}{n} \sum_{k:u<k/n<v} f_n\left(\frac{k}{n}\right)$$

$$= \int_u^v f_n(x)dx + o(1),$$

as $n \rightarrow \infty$. Hence,

$$P\left(u < \frac{L_{2n}}{2n} < v\right) \rightarrow \lim_{n\to\infty} \int_u^v f_n(x)dx = \int_u^v f(x)dx.$$

Integrating now leads to the desired result. $\square$

3.3   Exercises

Exercise 3.3.1. Consider a random walk S_i, which starts in 0, and makes $2n$ steps. Let $k < n$. Show that

(a) $E(|S_{2k}| \mid |S_{2k-1}| = r) = r$;

(b) $E(|S_{2k+1}| \mid |S_{2k}| = r) = 1$ if $r = 0$, and $E(|S_{2k+1}| \mid |S_{2k}| = r) = r$ otherwise.

Exercise 3.3.2. Consider a random walk making $2n$ steps, and let T be the first return to its starting point, that is

$$T = \min\{1 \leq k \leq 2n : S_k = 0\},$$

3.3. *Exercises*

and $T = 0$ if the walk does not return to zero in the first $2n$ steps. Show that for all $1 \leq k \leq n$ we have,

$$P(T = 2k) = \frac{1}{2k - 1} \binom{2k}{k} 2^{-2k}.$$

Exercise 3.3.3. Show that it already follows from Lemma 3.1.14 that

$$P(L_{2n} = 0) = P(L_{2n} = 2n).$$

Chapter 4

Limit Theorems

We have already encountered a number of limit theorems. For instance, the law of large numbers Theorem 1.6.1 and the arc-sine law Theorem 3.2.3 were statements about the limiting behaviour of a growing number of random variables. In this chapter, we state new laws of large numbers, and a primitive version of the famous central limit theorem.

The general set up in this section is as follows. We have random variables $X_1, \ldots, X_n$ defined on some sample space, with probability measure P_n. We typically do not express the dependence on n of this probability measure, and write P instead of P_n.

4.1 The Law of Large Numbers

The law of large numbers in Theorem 1.6.1 was for very special random variables, namely those which only take the values zero and one. Note that when X takes the values zero and one with equal probability (as in Theorem 1.6.1), the expectation of X is $1/2$. Hence Theorem 1.6.1 says something about convergence of averages to an *expectation*. In the meantime, that is, since Theorem 1.6.1, we have defined and discussed the expectation of general discrete random variables, and therefore it makes sense to try to state (and prove) a more general law of large numbers which should again say something about convergence of averages to an expectation. It turns out that when we assume that the variance is finite, this is in fact quite easy, using Chebyshev's inequality 2.3.27.

Theorem 4.1.1 (Weak law of large numbers). *Let $X_1, X_2, \ldots, X_n$ be independent random variables with the same distribution. Let $E(X_1) = \mu$ and $\mathrm{var}(X_1) = \sigma^2$ both be finite. Denote the sum $X_1 + \cdots + X_n$ by S_n. Then, for any $\epsilon > 0$ we have*

$$P\left(\left|\frac{S_n}{n} - \mu\right| > \epsilon\right) \leq \frac{\sigma^2}{n\epsilon^2},$$

which tends to zero when n tends to infinity.

Proof. We first compute the expectation and variance of S_n/n. The expectation of S_n/n is equal to μ, and to compute the variance, we write

$$\text{var}\left(\frac{S_n}{n}\right) = \text{var}\left(\frac{1}{n}\sum_{i=1}^{n}X_i\right) = \frac{1}{n^2}\sum_{i=1}^{n}\text{var}(X_i) = \frac{\sigma^2}{n}.$$

It then follows from Chebyshev's inequality 2.3.27 that

$$P\left(\left|\frac{S_n}{n} - \mu\right| > \epsilon\right) \leq \frac{\text{var}(S_n/n)}{\epsilon^2}$$

$$= \frac{\sigma^2}{n\epsilon^2} \to 0,$$

as n tends to infinity. □

♦ **Exercise 4.1.2.** Show that this law of large numbers generalises Theorem 1.6.1.

♦ **Exercise 4.1.3.** Relax the condition that all X_i's should have the same distribution. Can you also relax the assumption of independence? (Look at the proof what you really need to make it work.)

The weak law of large numbers as just stated is useful in many applications, we shall see examples of this later. However, one of the assumptions is that the variance of the random variables in question is finite. This condition is sometimes of course not met, and since the amount of extra work needed to get rid of this assumption is relatively modest, here follows a more general law of large numbers. An extra argument to include this one is that its proof is based on a very important technique in probability theory, namely the technique of *truncation*.

Before we start, let us recall one more piece of notation. When X is a random variable, then we can define new random variables like $\mathbf{1}_{\{X>a\}}$ which takes the value 1 if $X > a$ and the value 0 otherwise, that is,

$$\mathbf{1}_{\{X>a\}}(\omega) = \begin{cases} 1 & \text{if } X(\omega) > a \\ 0 & \text{if } X(\omega) \leq a. \end{cases}$$

The subscript can of course be any restriction on X. Note that the expection of such random variables is easily computed, since they take only two values. For instance, in the above case we have

$$E(\mathbf{1}_{\{X>a\}}) = P(X > a).$$

Theorem 4.1.4 (General weak law of large numbers). *Let $X_1, X_2, \ldots, X_n$ be independent and identically distributed random variables, with $E(|X_1|) < \infty$ and write $E(X_1) = \mu$. Then for any $\epsilon > 0$ we have*

$$P\left(\left|\frac{S_n}{n} - \mu\right| > \epsilon\right) \to 0,$$

as $n \to \infty$.

Before we prove this important result, we state a lemma which generalises Exercise 2.3.8.

Lemma 4.1.5. *Let X be a random variable taking only non-negative values, and let $p > 0$. Then*

$$E(X^p) = \int_0^\infty px^{p-1}P(X > x)dx.$$

Proof. We start with the right hand side and work our way to the left hand side as follows.

$$
\begin{aligned}
\int_0^\infty px^{p-1}P(X > x)dx &= \int_0^\infty px^{p-1}\sum_k P(X = k)\mathbf{1}_{\{k>x\}}(x)dx \\
&= \sum_k P(X = k)\int_0^\infty px^{p-1}\mathbf{1}_{\{x<k\}}(x)dx \\
&= \sum_k P(X = k)\int_0^k px^{p-1}dx \\
&= \sum_k P(X = k)k^p = E(X^p).
\end{aligned}
$$

$\square$

Proof of Theorem 4.1.4. First, we observe that for $x > 0$ we have

$$
\begin{aligned}
xP(|X_1| > x) &= xE(\mathbf{1}_{\{|X_1|>x\}}) \le E(|X_1|\mathbf{1}_{\{|X_1|>x\}}) \\
&= \sum_{k>x} kP(|X_1| = k) \\
&\to 0,
\end{aligned}
$$

when $x \to \infty$ since $E(|X_1|) = \sum_k kP(|X_1| = k) < \infty$ by assumption.
Next, we define

$$\mu_n = E(X_1\mathbf{1}_{\{|X_1|\le n\}}) = \sum_{k\le n} kP(X_1 = k).$$

Since $E(X_1) < \infty$, we have that $\mu_n \to E(X_1) = \mu$, as $n \to \infty$. This means that it suffices to show that (writing S_n for the sum of the appropriate random variables as usual)

$$P\left(\left|\frac{S_n}{n} - \mu_n\right| > \epsilon\right) \to 0, \tag{4.1}$$

as $n \to \infty$. In order to do that we use truncation. We define new random variables $X_{n,k}$ as follows:

$$X_{n,k} = X_k\mathbf{1}_{\{|X_k|\le n\}}.$$

Finally, we set

$$S_n' = X_{n,1} + \cdots + X_{n,n}.$$

We may now write

$$P\left(\left|\frac{S_n}{n} - \mu_n\right| > \epsilon\right) = P\left(\left|\frac{S_n}{n} - \mu_n\right| > \epsilon, S_n = S_n'\right) +$$

$$+ P\left(\left|\frac{S_n}{n} - \mu_n\right| > \epsilon, S_n \neq S_n'\right)$$

$$\leq P\left(\left|\frac{S_n'}{n} - \mu_n\right| > \epsilon\right) + P\left(S_n \neq S_n'\right).$$

The second term at the right hand side is easily estimated, since

$$P\left(S_n \neq S_n'\right) \leq P\left(X_k \neq X_{n,k} \text{ for some } k \leq n\right)$$
$$\leq nP\left(|X_1| > n\right),$$

which tends to zero, as observed at the beginning of this proof.

The first term on the right hand side is more complicated. From Markov's inequality 2.3.26 it follows that

$$P\left(\left|\frac{S_n'}{n} - \mu_n\right| > \epsilon\right) \leq \frac{E\left(\left|\frac{S_n'}{n} - \mu_n\right|^2\right)}{\epsilon^2}$$

$$= \frac{1}{n^2\epsilon^2} E\left(\sum_{i=1}^{n} X_{n,i} - n\mu\right)^2$$

$$= \frac{1}{n^2\epsilon^2} \operatorname{var}\left(\sum_{i=1}^{n} X_{n,i}\right)$$

$$= \frac{n\operatorname{var}(X_{n,1})}{\epsilon^2 n^2} \leq \frac{E(X_{n,1}^2)}{n\epsilon^2}.$$

According to Lemma 4.1.5 we have

$$E\left(\frac{X_{n,1}^2}{n}\right) = \frac{1}{n}\int_0^\infty 2xP(|X_{n,1}| > x)dx$$

$$\leq \frac{1}{n}\int_0^n 2xP(|X_1| > x)dx.$$

According to the first claim in this proof again, the integrand of the last integral converges to 0 as $x \to \infty$. Hence, for any small $\delta > 0$, we can find x_0 so that $2xP(|X_1| > x) < \delta$ for all $x > x_0$. Writing $M < \infty$ for the supremum over all x of the integrand, we can then bound the right hand side by

$$\frac{M}{n}x_0 + \delta,$$

and the proof is complete.   □

Example 4.1.6. When you play roulette in a casino, the probability to lose is always slightly larger than the probability to win. Suppose for simplicity that you win one euro with probability 49/100 and lose one with probability 51/100. We can model this by introducing independent random variables $X_1, X_2, \ldots$ with $P(X_i = 1) = 49/100$ and $P(X_i = -1) = 51/100$. A short computation shows that $E(X_i) = -1/50$. Your capital after n games is $S_n = X_1 + \cdots + X_n$. The law of large numbers now tells you that

$$P\left(\left| \frac{S_n}{n} + \frac{1}{50} \right| > \epsilon \right) \to 0,$$

as $n \to \infty$. Hence S_n/n will, with high probability, be close to -1/50, and therefore S_n will be close to $-n/50$. You are bound to lose. Also, the probability that after n games you have won, can be estimated as

$$
\begin{aligned}
P(S_n > 0) &= P\left(\frac{S_n}{n} > 0 \right) \\
&\leq P\left(\left| \frac{S_n}{n} + \frac{1}{50} \right| > \frac{1}{50} \right),
\end{aligned}
$$

which tends to 0 as $n \to \infty$, according to the law of large numbers. $\qquad\square$

4.2 The Central Limit Theorem

The law of large numbers tells us that the average of a number of random variables is in soms sense close to the expectation of any of the individual random variables. But how close is close? All we know from the law of large numbers, is that when we divide $S_n = X_1 + \cdots + X_n$ by n, the result will, with high probability at least, be near $E(X_1)$. But dividing by n is in some sense quite coarse. Deviations which are typically smaller than n cannot be seen anymore after dividing by n. For instance, for all we know, S_n could be near $E(X_1) \pm \log n$ or near $E(X_1) \pm \sqrt{n}$, or $\ldots$

The central limit theorem will tell us that S_n will typically deviate by a multiplicative factor of $\sqrt{n}$ from the expectation. This will be true for *any* distribution of the random variables $X_1, \ldots, X_n$, as long as they have finite variance. In this section, we will state and prove a special case of this theorem in the case that the X_i's take only the values 1 and -1 with equal probability. For this special case we can give a proof based on counting, very much in the same spirit as the proof of the arc-sine law in the previous chapter.

Theorem 4.2.1 (Central limit theorem). *Let $X_1, X_2, \ldots, X_n$ be independent random variables with the same distribution, taking the values ± 1 with equal probability. Let $S_n = X_1 + \cdots + X_n$ be their sum. Then, for any $a < b$ we have*

$$P\left(a \leq \frac{S_n}{\sqrt{n}} \leq b \right) \to \int_a^b \frac{1}{\sqrt{2\pi}} e^{-x^2/2} dx,$$

as $n \to \infty$.

Proof. We have seen already in Chapter 3 that

$$P\left(S_{2n} = 2k\right) = \binom{2n}{n+k} 2^{-2n}. \tag{4.2}$$

As in the proof of the arc-sine law, we will now use Stirling's formula to estimate this probability. It will become clear from this where the somewhat magic factor $\sqrt{n}$ in the statement of the theorem comes from. Recall that Stirling's formula says that

$$n! \sim n^n e^{-n} \sqrt{2\pi n},$$

as $n \to \infty$. Applying this to (4.2) leads to

$$
\begin{aligned}
\binom{2n}{n+k} 2^{-2n} &= \frac{(2n)!}{(n+k)!(n-k)!} 2^{-2n} \\
&\sim \frac{(2n)^{2n} 2^{-2n}}{(n+k)^{n+k}(n-k)^{n-k}} \cdot \frac{(2\pi(2n))^{1/2}}{(2\pi(n+k))^{1/2}(2\pi(n-k))^{1/2}} \\
&= \left(\frac{n}{n+k}\right)^{n+k} \left(\frac{n}{n-k}\right)^{n-k} \\
&\quad \times \left(\frac{n}{n+k}\right)^{1/2} \left(\frac{n}{n-k}\right)^{1/2} \cdot (\pi n)^{-1/2} \\
&= \left(1+\frac{k}{n}\right)^{-n-k} \left(1-\frac{k}{n}\right)^{-n+k} \\
&\quad \times \left(1+\frac{k}{n}\right)^{-1/2} \left(1-\frac{k}{n}\right)^{-1/2} \cdot (\pi n)^{-1/2} \\
&= \left(1+\frac{k}{n}\right)^{-k} \left(1-\frac{k}{n}\right)^{k} \left(1-\frac{k^2}{n^2}\right)^{-n} \\
&\quad \times \left(1+\frac{k}{n}\right)^{-1/2} \left(1-\frac{k}{n}\right)^{-1/2} \cdot (\pi n)^{-1/2}
\end{aligned}
$$

This somewhat boring computation does in fact bring us to an important observation. At this point we need to use one of the standard limits from calculus, namely the fact that

$$\lim_{n\to\infty} \left(1+\frac{x}{n}\right)^n = e^x,$$

for all $x \in \mathbb{R}$. It is probably not immediately clear how we can use this in the computation above. In fact, we can only do this for a special choice of k, and we shall now choose a k which depends on n. For a fixed $x \in \mathbb{R}$ we choose $2k$ to be the nearest integer to $x\sqrt{2n}$. This means that $|2k - x\sqrt{2n}|$ is at most one. If we pretend that $2k$ is exactly equal to $x\sqrt{2n}$, we see that the third term can be written as

$$\left(1-\frac{k^2}{n^2}\right)^{-n} = \left(1-\frac{x^2}{2n}\right)^{-n} \to e^{x^2/2},$$

as $n \to \infty$. Similarly, we find

$$
\left(1 + \frac{k}{n}\right)^{-k} = \left(1 + \frac{x}{\sqrt{2n}}\right)^{-x\sqrt{n/2}}
$$

$$
= \left(1 + \frac{x^2/2}{x\sqrt{n/2}}\right)^{-x\sqrt{n/2}}
$$

$$
\to e^{-x^2/2},
$$

as $n \to \infty$. A similar computation shows that also

$$
\left(1 - \frac{k}{n}\right)^{k} \to e^{-x^2/2}.
$$

♠ **Exercise 4.2.2.** Give a proof of this last statement.

Finally, when $|2k - x\sqrt{2n}| \le 1$, the terms $(1 + k/n)^{-1/2}$ and $(1 - k/n)^{-1/2}$ both converge to 1, as $n \to \infty$. Putting all these things together then gives

$$
P(S_{2n} = 2k) \sim (\pi n)^{-1/2} e^{-x^2/2},
$$

when we let $n \to \infty$, and choose k according to $2k = x\sqrt{2n}$. At this stage of the proof, the magic $\sqrt{n}$ in the statement of the theorem becomes somewhat clearer. It is only when we choose k and n related via $2k = x\sqrt{2n}$ that we obtain a non-trivial limit here. In some sense, the computation forces us to make this choice.

Apparently, S_{2n} typically takes values of the order $\sqrt{2n}$. So we are now motivated to compute

$$
P(a\sqrt{2n} \le S_{2n} \le b\sqrt{2n}),
$$

for any $a < b$. To this end we write

$$
P(a\sqrt{2n} \le S_{2n} \le b\sqrt{2n}) = \sum_{m:a\sqrt{2n}\le 2m\le b\sqrt{2n}} P(S_{2n} = 2m)
$$

$$
= \sum_{m:a\le 2m/\sqrt{2n}\le b} P(S_{2n} = 2m)
$$

$$
\sim \sum_{a\le x\le b: x\in 2\mathbb{Z}/\sqrt{2n}} (\pi n)^{-1/2} e^{-x^2/2},
$$

where the last equality comes from the substitution $2m = x\sqrt{2n}$, and where $2\mathbb{Z}/\sqrt{2n}$ is the set $\{2z/\sqrt{2n} : z \in \mathbb{Z}\}$. Now the last sum is a Riemann sum with step size $(2/n)^{1/2}$. So we rewrite the last sum as

$$
\sum_{a\le x\le b: x\in 2\mathbb{Z}/\sqrt{2n}} (2\pi)^{-1/2} e^{-x^2/2} (2/n)^{1/2}.
$$

It is now clear that when $n \to \infty$, this converges to

$$\frac{1}{\sqrt{2\pi}} \int_a^b e^{-x^2/2} dx,$$

which is what we wanted to prove. □

♠ **Exercise 4.2.3.** The above proof is valid for even n. Extend the result to all n.

4.3 Exercises

Exercise 4.3.1. Consider the following situation. We have two collections of random variables X_i and Y_i. The X_i's are independent and identically distributed with expectation 1, and the Y_i's are also independent and identically distributed with expectation 2. Now we first flip a coin once, and we define, for all n, a random variable S_n by $S_n = \sum_{i=1}^n X_i$ if heads comes up, and $S_n = \sum_{i=1}^n Y_i$ if tail comes up. Show that

(a) $E(S_n/n) = 3/2$.

(b) $P(|S_n/n - 3/2| > 1/4)$ does *not* converge to 0, when $n \to \infty$.

(c) Why does (b) not contradict the law of large numbers?

Exercise 4.3.2 (Local central limit theorem). Let S_n have a Poisson distribution with parameter n. Use Stirling's formula to show that when $(k-n)/\sqrt{n} \to x$, then

$$\sqrt{2\pi n} P(S_n = k) \to e^{-x^2/2}.$$

Chapter I

Intermezzo

So far we have been dealing with countable sample spaces. The reason for this, as mentioned before, is that we can develop many interesting probabilistic notions without needing too many technical details. However, there are a number of intuitive probabilistic ideas which can not be studied using only countable sample spaces. Perhaps the most obvious kind of things which can not be captured in the countable setting are those involving an *infinitely fine* operation, such as choosing a random point from a line segment. In this intermezzo we see why this is so, and observe some problems that arise from this fact. After the intermezzo, we continue with the study of continuous probability theory, which does capture such infinitely fine operations.

I.1 Uncountable Sample Spaces

When the sample space is countable, any event A clearly has only countably many elements. Hence the sum

$$P(A) = \sum_{\omega \in \Omega} p(\omega), \tag{I.1}$$

which describes the probability of an event A, has a definite meaning. This is no longer true when the sample space is not countable, since in that case we cannot list all elements, and we often cannot write down a sum as in (I.1).

Perhaps you wonder what kind of sample spaces are not countable. Here is a major example.

Example I.1.1 (Picking a random point from a line segment). Suppose we want to make a model which describes the idea of picking a point randomly from the unit interval $[0, 1]$. The natural sample space is simply $\Omega = [0, 1]$, and it is clear that this is an infinite space. Is Ω countable? The answer is no, and this can be seen as follows.

All points $\omega \in \Omega$ have a so called *binary expansion*. This expansion is defined as follows. We can write

$$\omega = \sum_{n=1}^{\infty} \frac{d_n(\omega)}{2^n},$$

which we can also write as

$$\omega = .d_1(\omega)d_2(\omega)\cdots,$$

where $d_n(\omega)$ is 0 or 1. For some ω, this expansion is not unique, for instance, $1/2 = .1000\ldots$ and $1/2 = .0111\ldots$. But for all ωs which are not a multiple of 2^{-n} for some n, there is a unique binary expansion.

Now suppose that $[0,1]$ is countable. Each $\omega \in [0,1]$ has a binary expansion, and some will have two such expansions, as observed above. Hence the collection of binary expansions arising from $[0,1]$ is countable: we simply consider the elements of $[0,1]$ one by one, and write down the one or two corresponding binary expansions.

Let us denote the collection of binary expansions by $\omega^1, \omega^2, \ldots$, so that each ω^i is in fact an infinite sequence of 0s and 1s:

$$\omega^i = (\omega_1^i, \omega_2^i, \omega_3^i, \ldots).$$

Now we define a special sequence ω^* as follows:

$$\omega^* = (1 - \omega_1^1, 1 - \omega_2^2, 1 - \omega_3^3, \ldots).$$

This looks perhaps a little weird, but the only thing we do is to take the nth element of ω^n, and put 1 minus this nth element on the nth position of ω^*. Now have a careful look at the newly constructed ω^*. Since $\omega_1^* = 1 - \omega_1^1 \neq \omega_1^1$, it is clear that $\omega^* \neq \omega^1$. Actually, since $\omega_n^* = 1 - \omega_n^n$, we see that the nth coordinates of ω^* and ω^n are different. Hence $\omega^* \neq \omega^n$ for all $n = 1, 2, \ldots$ But now we have reached a contradiction: we have found a binary expansion which is not one of our original ω^n's. Hence our original assumption is wrong, and we conclude that $[0,1]$ is not countable. $\qquad\square$

A set which is not countable will be said to be *uncountable*. Hence we have proved that $[0,1]$ is uncountable. This means that the theory of the preceding chapters can not be used. In the next section we shall see that this has serious consequences.

I.2 An Event Without a Probability?!

Perhaps the fact that certain sample spaces are not countable will not really terrify you. Of course, we can no longer use the idea of expressing probabilities as the sum of the appropriate probability masses, but perhaps there is a better, or more general way of defining and computing probabilities. For instance, when you pick a random point from a line segment, you could start with the idea that the probability that this point ends up in a given interval, should depend only on the length of this interval.

As an example, consider picking a point randomly from $[0, 1]$. The probability that the chosen point lies in $[0, \frac{1}{4}]$ *should* be $\frac{1}{4}$. Indeed, the probability that the point ends up in any given interval of length r should be r.

More generally, one could hope for some way of assigning a 'length', 'area' or 'volume', $V(A)$ say, to any subset A of a sample space Ω, and then define the probability of A as

$$P(A) = \frac{V(A)}{V(\Omega)}.$$

The goal of the current section is to convince you that there is, in fact, a serious and fundamental problem with this approach. The following striking example tells us that there is no hope that such an approach could work without problems.

Example I.2.1 (Banach–Tarskii paradox). Suppose that we want to select a random point from the surface of the unit sphere in three dimensions. We have no machinery yet how to define a probability measure in this context, but it seems reasonable to do the following. For any subset A of the surface of the unit sphere, we would like to have the property that the probability that the chosen point is in A should be proportial to the area of A. In other words, if the set A becomes twice as big, the probability to find our chosen point in A should also become twice as big.

Agreeing on the reasonableness of this idea, we now turn to the Polish mathematicians Banach and Tarskii. They constructed a most remarkable subset of the surface of the unit sphere. This subset, we call it A, has the following property. One can find rotations A' and A'' of A on the surface of the sphere, in such a way that A, A' and A'' are pairwise disjoint, and in addition, such that $A \cup A' \cup A''$ consists of the whole surface. Since the areas of A, A' and A'' are all the same, it seems very reasonable, and inescapable, to assign probability $1/3$ to each of the sets. In particular, the probability that our chosen point falls in A should be equal to $1/3$.

So far so good, sets A, A' and A'' with these properties are easy to find. But now comes the unbelievable fact. It is also possible to find rotations B, B' and B'' of the *same* set A, in such a way that the collection A, B, B', B'' is pairwise disjoint and together covers the whole surface! Since these *four* sets are again rotations of each other, they should all have the same probability, and hence this probability must now be $1/4$. In particular, the probability that our chosen point falls in A should be $1/4$. This contradicts our earlier observation, and we are in deep trouble. □

It seems almost magic that a set A as in Example I.2.1 can possibly exist, but yet, it does. This very counterintuitive example shows that it may, after all, not be so easy to assign probabilities to *all* subsets of a given sample space. In the example, it is impossible to assign a probability to A if we insist on rotation invariance; the set A has no well defined area.

What to do? We shall resolve this problem by insisting that the collection of events which receive a probability has a certain structure; *not all sets will receive a probability*. This is the only possible solution, we have to restrict the collection of events.

I.3   Random Variables on Uncountable Sample Spaces

We end this first intermezzo with some more motivation for the next chapter. We have noted in this intermezzo that very natural sample spaces can be uncountable, and that in such an uncountable sample space, certain strange subsets can exist which we can not assign a probability. This leads to serious complications if we want to define random variables on such sample spaces. As in Chapter 2, we would like to define a random variable X as a function

$$X : \Omega \to \mathbb{R},$$

with the property that we can speak of the probability that X takes a certain value a, or the probability that $a < X < b$, for instance. Hence, we would like, for instance, to assign a well defined probability to sets of the form

$$\{\omega \in \Omega : a < X(\omega) < b\}.$$

But how do we know that this set in fact can be given a well defined probability? If the sample space Ω is uncountable this is not always clear, as was illustrated with the Banach–Tarskii paradox. This is one of the fundamental issues in probability theory.

The classical approach to attack this problem is to use measure theory to decide which sets receive a probability. But we write this book for students who haven't studied measure theory. Therefore, we follow a different approach, and only assign probabilities to sets for which certain Riemann integrals over these sets exist. This approach has the drawback that we have to restrict our sample spaces to $\mathbb{R}^d$, but in a first and second course in probability, this is no real restriction. At the end of the book, in Chapter 9, we will sketch the more general framework of probability theory provided by measure theory.

Chapter 5

Continuous Random Variables and Vectors

We have seen in the Intermezzo that there is a need to generalise the notions of an experiment and a random variable. In this chapter we suggest a set up which allows us to do this. As in the first two chapters, we first define experiments, and after that random variables. The theory in this chapter is built up similarly as the theory in the first two chapters.

5.1 Experiments

In the discrete theory, the sample space Ω was either finite or countably infinite. In the present context, this no longer suffices and we take $\Omega = \mathbb{R}^d$, that is, our experiments will have outcomes in $\mathbb{R}^d$, for some $d \in \mathbb{N}$.

An experiment in the discrete setting consisted of a sample space and a *probability mass function*, assigning a certain probability to each outcome in the sample space. The probability of an event A in the discrete setting could then simply be defined as the sum of all probabilities of elements in A. In the current continuous setting, this is impossible, but we can do something similar, replacing sums by appropriate integrals. Here are some examples of how this could work. After the examples, we give the formal definitions.

Example 5.1.1 (Choosing an arbitrary point from the unit interval). Suppose we want to model the choice of a completely random point in the interval $(0,1)$. There are uncountably many points in $(0,1)$, and hence we cannot list its elements. Instead of concentrating on the event that the chosen point is equal to a certain given element of $(0,1)$, we consider the event that the chosen point falls into a subinterval $I \subseteq (0,1)$. It is reasonable to assume that this probability should be equal to the length of I. Writing $|I|$ for the length of an interval I, we therefore *define* the probability of I as

$$P(I) = |I|.$$

You should think of this number as the probability that a completely random point in the unit interval ends up in I.

Another way of formulating this assignment of probabilities is the following. Define the function

$$f(x) = \begin{cases} 1 & \text{if } 0 < x < 1, \\ 0 & \text{elsewhere.} \end{cases}$$

Now observe that

$$P(I) = \int_I f(x)dx,$$

that is, we have written the probability of an interval I in terms of an integral. When we write things this way, there is no need to restrict ourselves to intervals contained in $(0, 1)$. Indeed, since $f(x) = 0$ for all $x \notin (0, 1)$, the part of an interval outside the unit interval does not contribute to the probability of the interval.

Formulated this way, we can also define the probability of other subsets of $\mathbb{R}$. Indeed, we can define

$$P(A) = \int_A f(x)dx, \tag{5.1}$$

as long as the integral $\int_A f(x)dx$ exists. This last requirement is necessary, since for instance

$$\int_{\mathbb{Q}} f(x)dx$$

does not exist, and therefore $P(\mathbb{Q})$ remains undefined. So not all subsets of $\mathbb{R}$ have a probability now, something we already anticipated in the Intermezzo.   □

When you compare (5.1) to the corresponding formula

$$P(A) = \sum_{x \in A} p(x)$$

in the discrete theory, then you see that the two expressions are very similar. The sum has been replaced by an integral, and the probability mass function has been replaced by the function f.

In the discrete theory, a different experiment required a different probability mass function. Here, in the current continuous setting, a different experiment requires a different function f. This is illustrated by the next example.

Example 5.1.2 (Darts). Suppose we play darts on a big circular board with radius 1 meter. Suppose that we are only interested in the distance between the arrow and the midpoint of this board, not in the exact position of the arrow.

What is the probability that the distance between the arrow and the midpoint is at most t, for $0 \le t \le 1$? To compute this, we assume that we hit the board in a completely random place and that we never miss the board. These assumptions imply that the probability to hit the board within the circle of radius t is the area of this smaller circle with radius t, divided by the area of the full circle with radius 1. These areas are πt^2 and π respectively, so the probability that the arrow is less than t away from the midpoint is

$$\frac{\pi t^2}{\pi} = t^2.$$

This means that the outcome of our experiment, that is, the distance from the arrow to the midpoint, is between 0 and t with probability t^2, and we can write this in integral form as follows:

$$P((0,t)) = t^2 = \int_0^t 2x\,dx,$$

for all $0 \le t \le 1$. It is reasonable now to define

$$g(x) = \begin{cases} 2x & \text{if } 0 < x < 1, \\ 0 & \text{elsewhere,} \end{cases}$$

and define

$$P(A) = \int_A g(x)\,dx,$$

for all $A \subseteq \mathbb{R}$ for which this integral is defined. $\qquad\square$

In the previous example, we reduced the problem to one dimension by looking only at the distance between the arrow and the midpoint of the board. It is, however, also possible to study this example in two dimensions:

Example 5.1.3 (Darts revisited). Consider the set-up of the previous example, but now we are interested in the *exact* position of the arrow, not only in the distance to the midpoint. Then the probability that the arrow ends up in a given region A should be proportional to the area of A. We formalise this as follows. Since the area of the full circular board is π, we define

$$f(x,y) = \begin{cases} \frac{1}{\pi} & \text{if } x^2 + y^2 \le 1, \\ 0 & \text{elsewhere.} \end{cases}$$

Probabilities can now be defined as

$$P(A) = \int\int_A f(x,y)\,dx\,dy,$$

for all A for which this integral is defined. In fact, this assignment of probabilities corresponds to the experiment of choosing an arbitrary point in the disc with radius 1. $\qquad\square$

We are almost ready now for the definition of an experiment in the current continuous context. Before we give this definition, we first need to define the analogue of probability mass functions in the continuous setting. This analogue is called a *probability density function* and is defined as follows.

Definition 5.1.4. We call a non-negative function $f : \mathbb{R}^d \to \mathbb{R}$ a (d-dimensional) *probability density function* if

$$\int_{a_1}^{b_1} \cdots \int_{a_d}^{b_d} f(x_1, \ldots, x_d) dx_d \cdots dx_1$$

exists for all $a_1 \leq b_1, \ldots, a_d \leq b_d$, and

$$\int_{-\infty}^{\infty} \cdots \int_{-\infty}^{\infty} f(x_1, \ldots, x_d) dx_d \cdots dx_1 = 1.$$

Often, we will simply call this a *density*.

Definition 5.1.5. A (d-dimensional) *experiment* is a sample space $\Omega = \mathbb{R}^d$, together with a (d-dimensional) density f. For any $A \subseteq \Omega$ for which

$$\int \cdots \int_A f(x_1, \ldots, x_d) dx_1 \cdots dx_d$$

is defined, we define the *probability of A* by

$$P(A) = \int \cdots \int_A f(x_1, \ldots, x_d) dx_1 \cdots dx_d.$$

As before, P is called a *probability measure*, and any subset of Ω that has a probability, is called an *event*.

Note that according to this definition, not all subsets are events.

In principle, any positive function satisfying the defining properties of a density gives rise to an experiment. However, not all experiments that arise in this way are meaningful from a probabilistic point of view. Typically, a density is only relevant if it can be associated with a 'real' probabilistic context. For instance, we already saw that the density f given by $f(x) = 1$ for $0 < x < 1$ and $f(x) = 0$ elsewhere, corresponds to choosing a random point from the unit interval. The densities in the other examples also had an obvious probabilistic interpretation.

We conclude this section with one more example, the *exponential* density function.

Example 5.1.6 (Waiting for an event to happen). Let $\lambda > 0$ and consider the density function

$$f(x) = \begin{cases} \lambda e^{-\lambda x} & \text{if } x > 0, \\ 0 & \text{elsewhere.} \end{cases}$$

Since $\int_0^\infty \lambda e^{-\lambda x} dx = 1$, f is indeed a density. This density is very often used to model the waiting time for a certain event to happen, like the waiting time for the next customer to arrive in a shop, or the waiting time for the next earthquake in California. We now sketch the reason for this. The argument goes via a suitable discretisation of the time axis, as follows.

Suppose we flip a coin at time epochs $\epsilon, 2\epsilon, \ldots$, and that Y is the waiting time for the first head. In the above terminology, the event we are waiting for is the occurrence of heads. Then Y/ϵ has a geometric distribution:

$$P(Y > k\epsilon) = (1 - p)^k,$$

where p is the probability of heads at each coin flip. Recall that $E(Y/\epsilon) = p^{-1}$, and therefore

$$E(Y) = \frac{\epsilon}{p}.$$

Now fix a time $t > 0$, and let $\epsilon \downarrow 0$. If we would keep the same p, then clearly the number of successes by time t would go to infinity, since the number of trials is roughly t/ϵ. So in order to compensate for the fact that the number of trials up to time t goes to infinity, we choose the success probability p in such a way that the expectation of Y is independent of the discrete time steps ϵ. More precisely, we choose p as to satisfy $p/\epsilon = \lambda$. Then we can write

$$
\begin{aligned}
P(Y > t) &= (1 - p)^{t/\epsilon} \\
&= (1 - \lambda\epsilon)^{t/\epsilon} \\
&= \left(1 - \frac{\lambda t \epsilon}{t}\right)^{t/\epsilon} \rightarrow e^{-\lambda t},
\end{aligned}
$$

as $\epsilon \rightarrow 0$. Hence in the limit as $\epsilon \rightarrow 0$, the probability that the waiting time for the first success is at least t is given by $e^{-\lambda t}$.

We can now set up an experiment (with outcome the waiting time) such that the probability that this waiting time is at least t is equal to $e^{-\lambda t}$, for all $t > 0$. A little thought reveals that we can do this by choosing the exponential density function $f(x) = \lambda e^{-\lambda x}$, since

$$\int_t^\infty \lambda e^{-\lambda x} dx = e^{-\lambda t},$$

for all $t > 0$. The probability that the waiting time is between a and b, for instance, can now be computed as

$$\int_a^b \lambda e^{-\lambda x} dx.$$

5.2   Properties of Probability Measures

In the current continuous setting, probabilities are expressed in terms of Riemann integrals. It is perhaps a good idea to recall the dominated convergence theorem 3.2.4 in the higher dimensional setting, since we will use this result a number of times below.

Theorem 5.2.1 (Dominated convergence). *Let f, f_i, $i = 1, 2, \ldots$ be positive integrable functions on $\mathbb{R}^d$, with $f_i(x) \to f(x)$ as $i \to \infty$, for all $x = (x_1, \ldots, x_d)$. Suppose in addition that there exists a function g with $\int \cdots \int_{\mathbb{R}^d} g(x)dx < \infty$ such that $f_i(x) \leq g(x)$, for all x and all i. Then*

$$\lim_{i \to \infty} \int \cdots \int_{\mathbb{R}^d} f_i(x)dx = \int \cdots \int_{\mathbb{R}^d} f(x)dx.$$

In the following lemma we collect some properties of probability measures. In what follows, when we write $P(A)$, this implicitly means that the probability of A is defined, that is, $\int_A f(x)dx$ exists for the appropriate density f. Recall the definition of the *indicator function* of an event E:

$$1_E(x) = \begin{cases} 1 & \text{if } x \in E, \\ 0 & \text{if } x \notin E. \end{cases}$$

Lemma 5.2.2. *Consider an experiment with sample space $\mathbb{R}^d$ and density f.*

(a) *For events $A_1, A_2, \ldots$ which are pairwise disjoint (and for which $\cup_{i=1}^{\infty} A_i$ is an event) we have*

$$P\left(\bigcup_{i=1}^{\infty} A_i\right) = \sum_{i=1}^{\infty} P(A_i).$$

(b) $P(A^c) = 1 - P(A)$.

(c) *If $A \subseteq B$, then $P(A) \leq P(B)$.*

(d) $P(A \cup B) = P(A) + P(B) - P(A \cap B)$.

(e) $P(\Omega) = 1$ *and* $P(\emptyset) = 0$.

(f) *For events $A_1 \subseteq A_2 \subseteq \cdots$ (such that $A = \cup_{i=1}^{\infty} A_i$ is an event) we have*

$$P(A) = \lim_{i \to \infty} P(A_i).$$

Proof. For (a), observe that for disjoint events $A_1, \ldots, A_n$ we have

$$P\left(\bigcup_{i=1}^{n} A_i\right) = \sum_{i=1}^{n} P(A_i), \tag{5.2}$$

as follows from elementary additivity properties of the Riemann integral. Next, consider an infinite sequence $A_1, A_2, \ldots$ of disjoint events such that $\cup_{i=1}^{\infty} A_i$ is an event. Then we have

$$
\begin{aligned}
\lim_{n \to \infty} P\left(\bigcup_{i=1}^{n} A_i\right) &= \lim_{n \to \infty} \int_{\cup_{i=1}^{n} A_i} f(x) dx \\
&= \lim_{n \to \infty} \int_{\mathbb{R}^d} f(x) \mathbf{1}_{\cup_{i=1}^{n} A_i}(x) dx \\
&= \int_{\mathbb{R}^d} f(x) \mathbf{1}_{\cup_{i=1}^{\infty} A_i}(x) dx \\
&= P\left(\bigcup_{i=1}^{\infty} A_i\right),
\end{aligned}
$$

where the one but last equality follows from Theorem 5.2.1. Hence the result follows by sending n to infinity on both sides of equation (5.2).

For (b), we write $P(A) = \int_A f(x) dx = \int_{\mathbb{R}^d} f(x) \mathbf{1}_A(x) dx$, and similarly $P(A^c) = \int_{\mathbb{R}^d} f(x) \mathbf{1}_{A^c}(x) dx$. Hence,

$$
\begin{aligned}
P(A) + P(A^c) &= \int_{\mathbb{R}^d} f(x) \mathbf{1}_A(x) dx + \int_{\mathbb{R}^d} f(x) \mathbf{1}_{A^c}(x) dx \\
&= \int_{\mathbb{R}^d} f(x) (\mathbf{1}_A(x) + \mathbf{1}_{A^c}(x)) dx \\
&= \int_{\mathbb{R}^d} f(x) dx = 1.
\end{aligned}
$$

Properties (c)-(e) also follow from elementary properties of the Riemann integral; see the forthcoming Exercise 5.2.3.

For (f) we apply Theorem 5.2.1 as follows. Define

$$
h_i(x) = \mathbf{1}_{A_i}(x) f(x) \text{ and } h(x) = \mathbf{1}_A(x) f(x).
$$

The assumptions imply that

$$
0 \leq h_i(x) \uparrow h(x)
$$

as $i \to \infty$. Now $\int_{\mathbb{R}^d} h(x) dx < \infty$ and hence, according to Theorem 5.2.1 we have

$$
\int_{\mathbb{R}^d} h_i(x) dx \uparrow \int_{\mathbb{R}^d} h(x) dx.
$$

The result now follows, since $\int_{\mathbb{R}^d} h_i(x) dx = P(A_i)$ and $\int_{\mathbb{R}^d} h(x) dx = P(A)$. $\square$

♠ **Exercise 5.2.3.** Give the details of the proof of (c)-(e). Write down exactly what properties of the Riemann integral you use.

5.3   Continuous Random Variables

The motivation to introduce random variables is the same as in the discrete case: often, we are not really interested in the actual outcome of an experiment, but in some *function* of this outcome instead. We first give the formal definition of a (continuous) random variable, and illustrate this definition with a number of examples. Recall the fact that our sample spaces are $\Omega = \mathbb{R}^d$, for some integer d.

Definition 5.3.1. Consider an experiment with sample space Ω and probability density function f.

(a) A *random variable* X is a mapping from Ω to $\mathbb{R}$ such that sets of the form

$$\{\omega \in \Omega : a \leq X(\omega) \leq b\}, \{\omega \in \Omega : a \leq X(\omega) < b\}$$

$$\{\omega \in \Omega : a < X(\omega) \leq b\}, \{\omega \in \Omega : a < X(\omega) < b\}$$

are events, for all $-\infty \leq a \leq b \leq \infty$. This to say that the integral of f over any of these sets exists.

(b) A random variable X is called a *continuous random variable* with *density g* if g is a density function and

$$P(a \leq X \leq b) = P(\omega : a \leq X(\omega) \leq b) = \int_a^b g(x)dx, \tag{5.3}$$

for all $-\infty \leq a \leq b \leq \infty$.

(c) A random variable X is called a *discrete random variable* if there is a countable subset C of $\mathbb{R}$ with $P(X \in C) = 1$.

Let us first make a number of remarks about Definition 5.3.1.

1. The definition in (a) expresses the idea that we want to be able to talk about the probability that X takes values in intervals. So for instance, we want to be able to talk about the probability that X takes a value between 5 and 12. Elementary properties of the Riemann integral then imply that many other probabilities are also well defined, for instance the probability that X takes a value between either 6 and 7 or between 9 and 10.

2. For a *continuous* random variable X, the probability that X lies between a and b is specified through an integral. This implies that the probability that X takes a value in a certain interval does not change when we include or exclude endpoints of the interval. It also implies that for a continuous random variable X, $P(X = a) = 0$ for all a.

3. Note that a continuous random variable X does not have a unique density. For example, if we have a density of X, then we can change this density in a single point to obtain another density of X. We come back to this issue in a moment, after we have defined distribution functions.

4. Perhaps you wonder why continuous random variables with a given density g exist. In fact, this is easy to see. If we start with an experiment $\Omega = \mathbb{R}$ with density g and define X through

$$X(\omega) = \omega,$$

then X is a continuous random variable with density g. This construction is often used, but we shall see soon that interesting random variables are certainly not always constructed like this.

Definition 5.3.2. The function

$$F_X(x) = P(X \le x)$$

is called the *distribution function* of the random variable X.

The following lemma implies that in order to show that a certain random variable is continuous, it suffices to show that its distribution function has the right form. The point of the lemma is that we do not assume from the outset that X is a *continuous* random variable. Rather, this follows as a *conclusion*.

Lemma 5.3.3. *Suppose that X is a random variable whose distribution function can be written as*

$$F_X(x) = \int_{-\infty}^{x} f(y)\,dy,$$

for some density f. Then X is a continuous *random variable with density f.*

Proof. First we prove that $P(X = a) = 0$ for all a. To see this, note that by Lemma 5.2.2(a) we have that

$$P(X \le a) = P(X < a) + P(X = a)$$

and therefore,

$$P(X = a) = P(X \le a) - P(X < a). \tag{5.4}$$

Since

$$\{X < a\} = \bigcup_{n=1}^{\infty} \left\{ X \le a - \frac{1}{n} \right\},$$

it follows from Lemma 5.2.2(f) that

$$
\begin{aligned}
P(X < a) &= \lim_{n \to \infty} P(X \le a - 1/n) \\
&= \lim_{n \to \infty} F_X(a - 1/n) \\
&= F_X(a) = P(X \le a),
\end{aligned}
$$

where the one but last equality follows since F_X is a continuous function by assumption. It now follows from (5.4) that $P(X = a) = 0$.

Let $a \leq b$. Since we can write

$$\{\omega : X(\omega) \leq b\} = \{\omega : X(\omega) \leq a\} \cup \{\omega : a < X(\omega) \leq b\},$$

a *disjoint* union, we have

$$P(X \leq b) = P(X \leq a) + P(a < X \leq b).$$

This leads to

$$
\begin{aligned}
P(a < X \leq b) &= P(X \leq b) - P(X \leq a) \\
&= \int_{-\infty}^{b} f(x)dx - \int_{-\infty}^{a} f(x)dx \\
&= \int_{a}^{b} f(x)dx,
\end{aligned}
$$

as required. Since $P(X = a) = 0$ for all a, we can interchange $<$ and $\leq$ signs, and we have proved the lemma. $\square$

Asking for the *distribution* of a continuous random variable is asking for either its distribution function or its density.

Let X be a continuous random variable with distribution function F_X, and suppose that F_X is differentiable on (a, b). It follows from the fundamental theorem of calculus that

$$F_X(b) - F_X(a) = \int_{a}^{b} \frac{d}{dt} F_X(t)dt.$$

Since

$$P(a \leq X \leq b) = F_X(b) - F_X(a),$$

this shows that $f(t) = \frac{d}{dt}F_X(t)$ is a density of X on (a, b). Our distribution functions will always be differentiable with at most finitely many exceptional points. Hence we can assume that the density f_X and the distribution function F_X are related via

$$f_X(x) = \frac{d}{dx}F_X(x), \tag{5.5}$$

for all x, with at most finitely many exceptions. Here follows a number of examples.

Example 5.3.4 (Uniform distribution). The random variable X is said to have a *uniform distribution* on $[a, b]$ if its density is given by

$$f(x) = \frac{1}{b - a},$$

for $a \leq x \leq b$, and $f(x) = 0$ otherwise. A random variable with a uniform distribution on $[a, b]$ can be interpreted as the outcome of a completely random point from $[a, b]$. (The interval $[a, b]$ need not be closed. If X has a uniform distribution on (a, b), then $f(x) = 1/(b - a)$ for all $x \in (a, b)$. This difference is of little importance.)

The distribution function of X is given by

$$
F_X(x) = \begin{cases} 0 & \text{if } x < a, \\ \frac{x-a}{b-a} & \text{if } a \leq x \leq b, \\ 1 & \text{if } x > b. \end{cases}
$$

To see how we can compute probabilities now, take for instance $a = 0$ and $b = 2$, so that X has a uniform distribution on $(0, 2)$. We compute $P(X > \frac{1}{2})$ as follows:

$$
P\left(X > \frac{1}{2}\right) = \int_{\frac{1}{2}}^{2} f(x)dx
$$

$$
= \int_{\frac{1}{2}}^{2} \frac{1}{2}dx = \frac{3}{4}.
$$

$\square$

♠ **Exercise 5.3.5.** Check that $f_X(x) = \frac{d}{dx}F_X(x)$ for all but two values of x in this example.

Example 5.3.6 (Normal distribution). The random variable X has a *standard normal distribution* if its density is given by

$$
f(x) = \frac{1}{\sqrt{2\pi}}e^{-\frac{1}{2}x^2},
$$

for all $x \in \mathbb{R}$. The first thing to check is that this is indeed a density. The integrand cannot be integrated directly, but there is a nice trick from calculus which enables us to compute $\int_{-\infty}^{\infty} f(x)dx$. The trick consists of computing the *product* of two such integrals, and then use polar coordinates:

$$
\int_{y=-\infty}^{\infty} f(y)dy \int_{x=-\infty}^{\infty} f(x)dx = \frac{1}{2\pi} \int_{y=-\infty}^{\infty} \int_{x=-\infty}^{\infty} e^{-\frac{1}{2}(x^2+y^2)} dxdy
$$

$$
= \frac{1}{2\pi} \int_{\theta=0}^{2\pi} \int_{r=0}^{\infty} re^{-\frac{1}{2}r^2} drd\theta = 1.
$$

The random variable X is said to have a *normal distribution* with parameters $\mu \in \mathbb{R}$ and $\sigma^2 > 0$ if

$$
f(x) = \frac{1}{\sqrt{2\pi\sigma^2}}e^{-\frac{1}{2}\frac{(x-\mu)^2}{\sigma^2}},
$$

for all $x \in \mathbb{R}$. Note that for $\mu = 0$ and $\sigma^2 = 1$, this reduces to the density of a standard normally distributed random variable.

♠ **Exercise 5.3.7.** Show that this is a density, using the fact that the density of the standard normal distribution integrates to 1.

The reason that the normal distribution is very important became already apparent in the central limit Theorem 4.2.1. Although we didn't call it by name at that moment, the limit really is expressed as a probability in terms of a standard normal distribution. In the next chapter, a much more general statement will be proved, again involving the normal distribution. □

Example 5.3.8 (Exponential distribution). The random variable X has an *exponential distribution* with parameter $\lambda > 0$ if its density is given by

$$f(x) = \lambda e^{-\lambda x},$$

whenever $x > 0$, and $f(x) = 0$ for $x \leq 0$. We have come across this density already in Example 5.1.6, where it naturally appeared as the density corresponding to the waiting time for an unpredictable event. In that example, we first derived the distribution function of the waiting time. Indeed, we showed that in the limit for $\epsilon \to 0$, the probability that the waiting time is at least t is equal to $e^{-\lambda t}$. Hence, the distribution function of the waiting time is $F(t) = 1 - e^{-\lambda t}$, for $t > 0$. When we differentiate this, we find the density of the exponential distribution, in agreement with (5.5). □

Example 5.3.9 (Cauchy distribution). The random variable X has a *Cauchy distribution* if its density is given by

$$f(x) = \frac{1}{\pi} \frac{1}{1 + x^2},$$

for all $x \in \mathbb{R}$. Here is a fairly natural example where the Cauchy distribution arises.

Consider the sample space $\mathbb{R}$ with a uniform density on $(-\frac{\pi}{2}, \frac{\pi}{2})$. We select a point Θ in this sample space, and construct a random variable X as follows. We draw a half-line in the plane, starting in $(0, 1)$ in direction Θ downwards. More precisely, Θ is the angle between the line and the y-axis, where Θ itself is uniformly distributed on $(-\frac{\pi}{2}, \frac{\pi}{2})$.

Denote the intersection of the line with the x-axis by $(X, 0)$. A little thought reveals that $X = \tan \Theta$. We claim that X is a continuous random variable with a Cauchy distribution. To see this, first observe that for any $-\pi/2 < a < \pi/2$, we have

$$P(\Theta \leq a) = \frac{a - (-\pi/2)}{\pi} = \frac{1}{2} + \frac{a}{\pi}.$$

Hence,

$$
\begin{aligned}
P(X \leq x) &= P(\tan \Theta \leq x) \\
&= P(\Theta \leq \arctan x) = \frac{1}{2} + \frac{1}{\pi} \arctan x,
\end{aligned}
$$

and differentiation now leads to the desired result. This is an example of a random variable X which is not the identity map. □

Example 5.3.10. Consider the darts Examples 5.1.2 and 5.1.3. We can reformulate this now in terms of continuous random variables. Indeed, the sample space in Example 5.1.3 is $\Omega = \mathbb{R}^2$ and the density is $f(x,y) = 1/\pi$ when $x^2 + y^2 \leq 1$, and $f(x,y) = 0$ otherwise. We can define a random variable X on Ω by

$$X((x,y)) = \sqrt{x^2 + y^2}.$$

Then X represents the distance to the midpoint, and we have seen in Example 5.1.2 that X is a random variable with density $f(x) = 2x$, when $0 < x < 1$ and $f(x) = 0$ elsewhere. In fact, in Example 5.1.2 we first obtained the distribution function, and derived the density from this by, indeed, differentiation, in agreement with (5.5). $\qquad\square$

5.4   Expectation

The concepts discussed in the first four chapters all have a counterpart here. When comparing results and definitions with the discrete theory, the general rule of thumb is that we replace probability mass functions by densities, and sums by integrals. For the expectation of a random variable, this leads to the following definition. As usual, We say that the integral $\int_{-\infty}^{\infty} g(x)dx$ is *well defined* if its positive and negative parts are not both infinite with opposite sign.

Definition 5.4.1. The *expectation* $E(X)$ of a random variable X with density f is defined by

$$E(X) = \int_{-\infty}^{+\infty} xf(x)dx,$$

whenever this integral is well defined.

Maybe you are not convinced yet that this is the right definition of the expectation. So let us explain why this definition is reasonable. For simplicity, we assume that the continuous random variable X is bounded and non-negative: $0 \leq X < K$ for some integer K. The argument goes via an approximation of X, as follows.

Let, for any integer n, X_n be defined as

$$X_n = \frac{k}{n} \quad \text{if} \quad \frac{k}{n} \leq X < \frac{k+1}{n},$$

which implies that X_n is a discrete random variable. The expectation of X_n can be computed as follows:

$$E(X_n) \quad = \quad \sum_{k=0}^{nK-1} \frac{k}{n} P\left(X_n = \frac{k}{n}\right)$$

$$= \quad \sum_{k=0}^{nK-1} \frac{k}{n} P\left(\frac{k}{n} \leq X < \frac{k+1}{n}\right)$$

$$= \quad \sum_{k=0}^{nK-1} \frac{k}{n} \int_{k/n}^{(k+1)/n} f_X(x)dx$$

$$= \quad \int_0^K s_n(x)f_X(x)dx,$$

where

$$s_n(x) = \frac{k}{n} \quad \text{if} \quad \frac{k}{n} \leq x < \frac{k+1}{n}.$$

We then note that

$$\left|E(X_n) - \int_{-\infty}^{\infty} xf_X(x)dx\right| \quad = \quad \left|\int_{-\infty}^{\infty} \{s_n(x) - x\}f_X(x)dx\right|$$

$$\leq \quad \int_{-\infty}^{\infty} |s_n(x) - x| \, f_X(x)dx$$

$$\leq \quad \frac{1}{n}\int_{-\infty}^{\infty} f_X(x)dx$$

$$= \quad \frac{1}{n},$$

so that $E(X_n) \to \int_{-\infty}^{\infty} xf_X(x)dx$. Since $X_n \to X$ as $n \to \infty$ and $E(X_n) \to \int_{-\infty}^{\infty} xf_X(x)dx$, it is reasonable to define $E(X) = \int_{-\infty}^{\infty} xf_X(x)dx$.

The expectation of a random variable can take the values $\pm\infty$, we shall see examples of this.

Example 5.4.2 (Exponential distribution). For a random variable X with an exponential distribution with parameter λ we can compute the expectation as follows:

$$E(X) \quad = \quad \int_0^{\infty} \lambda x e^{-\lambda x}dx$$

$$= \quad 1/\lambda.$$

$\square$

Expectations of continuous random variables share the properties of their discrete counterparts. At this point, we only note the following. More properties follow in Section 5.8.

Theorem 5.4.3. *When $E(X)$ exists, we have $E(aX + b) = aE(X) + b$, for all $a, b \in \mathbb{R}$.*

Proof. Since we have defined the expectation via the density, we need to compute the density of $Y = aX + b$. To do this, we suppose first that $a > 0$ and write,

$$P(Y \leq y) = P\left(X \leq \frac{y-b}{a}\right)$$

$$= \int_{-\infty}^{(y-b)/a} f_X(x)dx$$

$$= \int_{-\infty}^{y} \frac{1}{a} f_X\left(\frac{u-b}{a}\right) du,$$

by the substitution $u = ax + b$. Hence, the density of Y evaluated at u is given by $\frac{1}{a} f_X(\frac{u-b}{a})$. This implies that the expectation of Y is equal to

$$E(aX + b) = \int_{-\infty}^{\infty} \frac{u}{a} f_X\left(\frac{u-b}{a}\right) du$$

$$= \int_{-\infty}^{\infty} \left(\frac{b}{a} + x\right) f_X(x)a dx$$

$$= \int_{-\infty}^{\infty} b f_X(x)dx + a \int_{-\infty}^{\infty} x f_X(x)dx$$

$$= b + aE(X),$$

by the substitution $u - b = xa$. $\qquad\square$

♠ **Exercise 5.4.4.** Give the analogous proof for $a < 0$.

Example 5.4.5 (Uniform distribution). Recall that a random variable X is said to have a *uniform distribution* on $[a, b]$ if its density is given by

$$f(x) = \frac{1}{b-a},$$

for $a \leq x \leq b$, and $f(x) = 0$ otherwise. Its expectation is given by

$$E(X) = \int_a^b \frac{x}{b-a} dx$$

$$= \frac{a+b}{2}. \qquad\square$$

Example 5.4.6 (Normal distribution). Recall that the random variable X has a *standard normal distribution* if its density is given by

$$f(x) = \frac{1}{\sqrt{2\pi}} e^{-\frac{1}{2}x^2},$$

for all $x \in \mathbb{R}$. It is easy to see, using the symmetry of f around 0, that the expectation of X is equal to 0. The random variable X was said to have a *normal distribution* with parameters $\mu \in \mathbb{R}$ and $\sigma^2 > 0$ if

$$f(x) = \frac{1}{\sqrt{2\pi\sigma^2}} e^{-\frac{1}{2}\frac{(x-\mu)^2}{\sigma^2}},$$

for all $x \in \mathbb{R}$. We can find the expectation of X via the following computation, writing $\sigma = \sqrt{\sigma^2}$. Let

$$Y = \frac{X - \mu}{\sigma}.$$

We claim that Y has a standard normal distribution. To see this, we can write

$$
\begin{aligned}
P(Y \leq y) &= P(X \leq y\sigma + \mu) \\
&= \frac{1}{\sigma\sqrt{2\pi}} \int_{-\infty}^{y\sigma+\mu} e^{-\frac{1}{2}\left(\frac{x-\mu}{\sigma}\right)^2} dx \\
&= \frac{1}{\sqrt{2\pi}} \int_{-\infty}^{y} e^{-\frac{1}{2}v^2} dv,
\end{aligned}
$$

where the last step follows from the substitution $v = (x - \mu)/\sigma$. This shows that Y has a standard normal distribution, so that $E(Y) = 0$. Since $X = \sigma Y + \mu$, this leads via Theorem 5.4.3 to $E(X) = \mu$. □

Example 5.4.7 (Cauchy distribution). Recall that the random variable X has a *Cauchy distribution* if its density is given by

$$f(x) = \frac{1}{\pi}\frac{1}{1 + x^2},$$

for all $x \in \mathbb{R}$. We claim that X has no expectation. The reason for this is that

$$\frac{1}{\pi}\int_{-\infty}^{\infty} \frac{x}{1 + x^2} dx$$

does not exist. (Can you verify this?) □

   We end this section with an envelope problem where continuous random variables play an important role.

Example 5.4.8 (Second envelope problem). Suppose that I show you two envelopes, both containing a certain amount of money. You have no information whatsoever about the amounts of money in the envelopes. (This is different from the first envelope problem in Example 2.5.10, where you had the information that one envelope contained twice the amount of the other.) You choose one and this time you do open the envelope. Suppose you see that the envelope contains x euros, say. After having seen this amount, I offer you the opportunity to swap, that is, to choose the other envelope. Does it make sense to do so? More precisely, is there a decision algorithm that enables you to end up with the highest amount of money, with probability strictly larger than $\frac{1}{2}$? It is perhaps surprising that such an algorithm does exist, and we describe it here now.

Suppose that the amounts in the two envelopes are a and b, with $a < b$. For the moment we assume that a and b are non-random. We denote the amount in the chosen envelope by X_1. This means that $P(X_1 = a) = P(X_1 = b) = \frac{1}{2}$. The algorithm makes use of an auxiliary random variable Y, which is independent of X_1 and which has an exponential distribution with parameter 1 say. (In fact, the precise distribution of Y is of little importance, we will see in a moment what properties this distribution should have for our purposes.) The algorithm runs as follows. After looking at the amount X_1, we go to our computer, and draw a realisation from Y. If $X_1 < Y$ then we swap, if $X_1 > Y$ then we don't. Quite simple, isn't it? Why does this work? Let us denote the amount of money in the final envelope (after possible swapping) by X_2. We can write down the joint distribution of (X_1, X_2). Indeed, $P(X_2 = b | X_1 = b) = P(Y < b) = 1 - e^{-b}$, and $P(X_2 = b | X_1 = a) = P(Y > a) = e^{-a}$. This leads to

$$
\begin{aligned}
P(X_2 = b) &= P(X_2 = b | X_1 = b) P(X_1 = b) + P(X_2 = b | X_1 = a) P(X_1 = a) \\
&= \frac{1}{2}(1 - e^{-b}) + \frac{1}{2}e^{-a} \\
&= \frac{1}{2} + \frac{1}{2}(e^{-a} - e^{-b}) > \frac{1}{2}.
\end{aligned}
$$

This is a rather surprising conclusion: there is a winning strategy in this case. □

♠ **Exercise 5.4.9.** Investigate what happens when we assume that the amounts in the two envelopes are themselves random variables.

♠ **Exercise 5.4.10.** Suppose that we have no information at all about the amounts in the two envelopes. In particular, we do not know whether the amounts are random, independent, etcetera. What kind of distributions of Y would be useful for the decision making procedure?

5.5   Random Vectors and Independence

As in the theory of discrete random variables, if we want to study the interaction between various random variables, we need to look at *random vectors*.

Definition 5.5.1. Consider an experiment with sample space Ω and density f.
(a) A function $X = (X_1, \ldots, X_d)$ from Ω into $\mathbb{R}^d$ is called a *random vector* if

$$
\{\omega : a_i \leq X_i(\omega) \leq b_i, i = 1, \ldots, d\}
$$

is an event, for all $-\infty \leq a_i \leq b_i \leq \infty$. This is to say that the integral of f over this set exists.

(b) A random vector $X = (X_1, \ldots, X_d)$ is called a *continuous random vector* with *(joint) density* g if

$$P(a_1 \leq X_1 \leq b_1, \ldots, a_d \leq X_d \leq b_d) = \int_{a_1}^{b_1} \cdots \int_{a_d}^{b_d} g(x_1, \ldots, x_d) dx_d \cdots dx_1,$$

$$(5.6)$$

for all $-\infty \leq a_i \leq b_i \leq \infty$, $i = 1, \ldots, d$.

As in the case of continuous random variables, once we have that $(X_1, \ldots, X_d)$ is a continuous random vector, many more probabilities of the form

$$P((X_1, \ldots, X_d) \in A)$$

are well defined, and equal to $\int_A f(x) dx$; this follows from elementary properties of the Riemann integral. In concrete cases, this will be understood without further comments.

Definition 5.5.2. We define the *(joint) distribution function* of $X = (X_1, \ldots, X_d)$ by

$$F_X(x_1, \ldots, x_d) = P(X_1 \leq x_1, \ldots, X_d \leq x_d).$$

We refer to the distribution of the vector as the *joint* ditribution, and to the distributions of its individual components as the *marginal* distributions. As in the discrete case, knowing the joint distribution means also knowing the marginal distribution of each of the individual X_i's:

Theorem 5.5.3. *Let* $X = (X_1, \ldots, X_d)$ *have joint distribution function* F_X *and joint density* f_X. *Then* X_i *is a continuous random variable with density*

$$f_{X_i}(x_i) = \int_{x_1=-\infty}^{\infty} \cdots \int_{x_{i-1}=-\infty}^{\infty} \int_{x_{i+1}=-\infty}^{\infty} \cdots$$
$$\cdots \int_{x_d=-\infty}^{\infty} f(x_1, \ldots, x_d) dx_d \cdots dx_{i+1} dx_{i-1} \cdots dx_1.$$

In words, we find the marginal density of X_i *by integrating out all other variables from the joint density.*

Proof. It follows from the definition that

$$P(X_i \leq x) = \int \cdots \int_{\{(x_1, \ldots, x_d): x_i \leq x\}} f_X(x_1, \ldots, x_d) dx_1 \cdots dx_d$$
$$= \int_{-\infty}^{x} \left(\int_{\infty}^{\infty} \cdots \int_{-\infty}^{\infty} f(x_1, \ldots, x_d) dx_1 \cdots dx_{i-1} dx_{i+1} \cdots dx_d \right) dx_i,$$

from which the result follows, using Lemma 5.3.3.   □

Example 5.5.4. Let (X, Y) have joint density

$$f(x, y) = \frac{1}{y} e^{-y - \frac{x}{y}},$$

for $x, y > 0$, and $f(x, y) = 0$ otherwise. The marginal density of Y can now be found as follows:

$$
\begin{aligned}
f_Y(y) &= \int_{-\infty}^{\infty} f(x, y) dx \\
&= \int_0^{\infty} \frac{1}{y} e^{-y - \frac{x}{y}} dx = e^{-y},
\end{aligned}
$$

for all $y > 0$. We conclude that Y has an exponential distribution with parameter 1. □

Example 5.5.5. Let (X, Y) have joint density

$$f(x, y) = \frac{1}{x}, \, 0 < y \le x \le 1,$$

and $f(x, y) = 0$ elsewhere. We find the marginal density of X by integrating out y:

$$f_X(x) = \int_{y=0}^{x} f(x, y) dy = 1,$$

where $0 < x \le 1$. Hence X has a uniform distribution on $(0, 1]$. The density of Y is equal to

$$f_Y(y) = \int_{x=y}^{1} f(x, y) dx = -\log y,$$

for $0 < y \le 1$. We have no particular name for this distribution. □

Recall that discrete random variables X and Y were said to be independent if

$$P(X = x, Y = y) = P(X = x)P(Y = y). \tag{5.7}$$

This definition is useless for continuous random variables. To see this, recall that for any continuous random variable X we have

$$P(X = x) = 0,$$

for all x. Hence (5.7) says that $0 = 0$, and this is a useless condition.

In the continuous context, there are several (equivalent) ways to define independence. we make a somewhat arbitrary choice, but note the forthcoming Exercise 5.5.8. Compare the following definition to Exercise 2.2.7.

Definition 5.5.6. The random variables X and Y, defined on the same sample space, are said to be *independent* if for all x and y,

$$P(X \le x, Y \le y) = P(X \le x)P(Y \le y).$$

♠ **Exercise 5.5.7.** State a suitable definition of independence for n random variables $X_1, \ldots, X_n$, defined on the same sample space.

♠ **Exercise 5.5.8.** Show that the continuous random variables X and Y with joint density f are independent if and only if there are densities f_X of X and f_Y of Y so that

$$f_X(x)f_Y(y) = f(x,y),$$

for all $x, y \in \mathbb{R}$.

Lemma 5.5.9. *The continuous random variables X and Y are independent if and only if they have a joint density $f(x,y)$ which can be written as a product $f(x,y) = g(x)h(y)$ of a function of x alone and a function of y alone. (We say, as before, that f factorises.)*

♠ **Exercise 5.5.10.** Prove this lemma. You may want to look at the proof of Theorem 2.4.8.

Example 5.5.11. Suppose that X and Y have joint density $f(x,y) = e^{-x-y}$, for $x, y > 0$, and $f(x,y) = 0$ elsewhere. Then $f(x,y)$ factorises as

$$f(x,y) = e^{-x}e^{-y},$$

and we conclude that X and Y are independent. □

Example 5.5.12. Consider Example 5.5.5. Note that X and Y are *not* independent, although it appears that the joint density is a product of a function of x alone and a function of y alone. Do you see why? □

Example 5.5.13 (Bivariate normal distribution). Suppose that (X, Y) is a random vector with density

$$f(x,y) = \frac{1}{2\pi\sqrt{(1-\rho^2)}}e^{-\frac{1}{2(1-\rho^2)}(x^2-2\rho xy+y^2)},$$

where ρ is a parameter taking values strictly between -1 and 1. This expression looks rather obscure and unattractive, but the point of this density will become clear now.

First, we need to convince ourselves that f is a density function, that is, we need to verify that $\int\int f(x,y)dxdy = 1$. We do this with a trick: we use our knowledge of one-dimensional normal distributions. A little algebra shows that

$$
\begin{aligned}
f(x,y) &= \frac{1}{2\pi\sqrt{(1-\rho^2)}}e^{-\frac{1}{2(1-\rho^2)}\left((x-\rho y)^2+y^2-\rho^2 y^2\right)} \\
&= \frac{1}{\sqrt{2\pi(1-\rho^2)}}e^{-\frac{1}{2(1-\rho^2)}(x-\rho y)^2} \cdot \frac{1}{\sqrt{2\pi}}e^{-\frac{1}{2}y^2}.
\end{aligned}
$$

We now integrate the right hand side first over x and then over y. Integration over x leaves the second term unchanged, and in the first term we recognise the density of a normal random variable with parameters ρy and $1 - \rho^2$. Hence when we integrate over x, the first term gives just 1, and we are left with the second term. However, the second term is the density of a standard normally distributed random variable, and therefore also integrates to 1 (integration is over y this time). Therefore, we find that $\int_y \int_x f(x, y) dx dy = 1$ as required.

Next, we want to compute the marginal distribution of X and Y. Well, this we have, in fact, already done in the previous calculation. Indeed, we showed that when we integrate out x, the result is the density of a standard normal distribution, and therefore, Y has a standard normal distribution. Since $f(x, y)$ is symmetric in x and y, it follows immediately that also X has a standard normal distribution. Note that this is true for all choices of the parameter ρ, and this is a very natural example in which the same marginal distributions have a different joint distribution. As in the discrete setting, the joint distribution does determine the marginals, but not the other way around. $\qquad \square$

5.6 Functions of Random Variables and Vectors

Given a random variable X or a random vector (X, Y), and an appropriate function g, when is $g(X)$ or $g(X, Y)$ a (continuous or discrete) random variable or vector? For instance, is $X + Y$ a random variable, when X and Y are?

First let's have a look at the one-dimensional case. Let X have density f. If we want to compute the distribution function of $g(X)$ we would like to write

$$P(g(X) \leq y) = P(X \in g^{-1}(-\infty, y]). \tag{5.8}$$

For this to make sense in our present context, $\{X \in g^{-1}(-\infty, y]\}$ must be a set which has received a probability. More generally, we want sets of the form $\{X \in g^{-1}(a, b)\}$ to have a well defined probability for all $a \leq b$. This is typically the case, and we illustrate this with some examples.

Example 5.6.1. Let X be a continuous random variable with differentiable distribution function F_X and density $f_X = \frac{d}{dx} F_X(x)$. Let $g(x) = 2x + 3$. Then

$$
\begin{aligned}
P(a \leq g(X) \leq b) &= P(a \leq 2X + 3 \leq b) \\
&= P\left(\frac{a-3}{2} \leq X \leq \frac{b-3}{2}\right) \\
&= \int_{(a-3)/2}^{(b-3)/2} f_X(x) dx \\
&= \int_a^b f_X\left(\frac{y-3}{2}\right) \frac{1}{2} dy,
\end{aligned}
$$

which means that $g(X)$ is a continuous random variable with $f_{g(X)}(y) = \frac{1}{2} f_X((y-3)/2)$. $\qquad \square$

Example 5.6.2. Let X have a standard normal distribution, and let $g(x) = x^2$. Then, writing $\Phi(x)$ for the distribution function of X, we have for $0 \leq x \leq y$,

$$
\begin{aligned}
P(x \leq g(X) \leq y) &= P(\sqrt{x} \leq X \leq \sqrt{y}) + P(-\sqrt{y} \leq X \leq -\sqrt{x}) \\
&= \Phi(\sqrt{y}) - \Phi(\sqrt{x}) + \Phi(-\sqrt{x}) - \Phi(-\sqrt{y}) \\
&= 2\Phi(\sqrt{y}) - 2\Phi(\sqrt{x}),
\end{aligned}
$$

since $\Phi(x) = 1 - \Phi(-x)$. Since $P(g(X) \leq y) = P(0 \leq g(X) \leq y)$, differentiating leads to

$$
f_{g(X)}(y) = 2\frac{d}{dy}\Phi(\sqrt{y}) = \frac{1}{\sqrt{2\pi y}}e^{-\frac{1}{2}y}.
$$

Hence $g(X)$ is a continuous random variable, with density

$$
f_{g(X)}(y) = \frac{1}{\sqrt{2\pi y}}e^{-\frac{1}{2}y},
$$

for all $y > 0$, and $f_{g(X)}(y) = 0$ for all $y \leq 0$. $\square$

In most cases of interest, it is possible to do a similar computation. It is also possible to write down a few general results, which reduce the amount of work in many cases.

Theorem 5.6.3. *Let X have density f, and let $g : \mathbb{R} \to \mathbb{R}$ be one-to-one and differentiable, with differentiable inverse. Then $g(X)$ is a continuous random variable with density*

$$
f_{g(X)}(y) = f(g^{-1}(y))\left|\frac{d}{dy}g^{-1}(y)\right|,
$$

for all y in the range of g, and $f_{g(X)}(y) = 0$ elsewhere.

Proof. Without loss of generality, we assume that g is non-decreasing. The result is a consequence of the classical change of variable theorem from calculus. Using the change of variables $z = g(x)$ we can write for $a \leq b$ such that (a, b) is in the range of g;

$$
\begin{aligned}
P(a \leq g(X) \leq b) &= P(g^{-1}(a) \leq X \leq g^{-1}(b)) \\
&= \int_{g^{-1}(a)}^{g^{-1}(b)} f(x)dx \\
&= \int_a^b f(g^{-1}(z))\left|\frac{d}{dz}g^{-1}(z)\right|dz,
\end{aligned}
$$

proving the result. $\square$

Example 5.6.4. Let X and g be as in Example 5.6.1. We can rederive the result from Example 5.6.1, using Theorem 5.6.3 as follows. The inverse $g^{-1}(y)$ is given by

$$
g^{-1}(y) = \frac{y - 3}{2},
$$

and hence $\frac{d}{dy}g^{-1}(y) = \frac{1}{2}$. Substituting this in Theorem 5.6.3 gives the result obtained in Example 5.6.1. $\qquad\square$

For the higher-dimensional case, we state the next theorem in two dimensions.

Theorem 5.6.5. *Let* (X_1, X_2) *have joint density* f, *and let* $g : \mathbb{R}^2 \to \mathbb{R}^2$ *be one-to-one, and write* $g(x_1, x_2) = (g_1(x_1, x_2), g_2(x_1, x_2)) = (y_1, y_2)$. *Since* g *is one-to-one, it can be inverted as* $x_1 = x_1(y_1, y_2)$ *and* $x_2 = x_2(y_1, y_2)$. *Let* J *be the Jacobian of this inverse transformation (where we assume enough differentiability). That is,*

$$J(y_1, y_2) = \frac{\partial x_1}{\partial y_1}\frac{\partial x_2}{\partial y_2} - \frac{\partial x_1}{\partial y_2}\frac{\partial x_2}{\partial y_1}.$$

Then $(Y_1, Y_2) = g(X_1, X_2)$ *is a continuous random vector with joint density*

$$f_{(Y_1, Y_2)}(y_1, y_2) = f(x_1(y_1, y_2), x_2(y_1, y_2))|J(y_1, y_2)|,$$

if (y_1, y_2) *is in the range of* g, *and* $f_{(Y_1, Y_2)}(y_1, y_2) = 0$ *otherwise.*

Proof. The proof is similar to the one-dimensional case and follows from the classical change of variables formula from calculus. $\qquad\square$

Example 5.6.6. Suppose that X_1 and X_2 are independent, exponentially distributed random variables with the same parameter λ. We compute the joint density of (Y_1, Y_2) where

$$Y_1 = X_1 + X_2 \text{ and } Y_2 = X_1/X_2.$$

To find this joint density, let g be the map defined by $g(x_1, x_2) = (x_1 + x_2, x_1/x_2) = (y_1, y_2)$. The inverse map is then given by

$$x_1 = \frac{y_1 y_2}{1 + y_2} \text{ and } x_2 = \frac{y_1}{1 + y_2},$$

and a simple computation shows that the Jacobian is equal to

$$J(y_1, y_2) = -y_1/(1 + y_2)^2.$$

Substituting this in the statement of the theorem gives

$$\begin{aligned} f_{(Y_1, Y_2)}(y_1, y_2) &= f_{(X_1, X_2)}\left(\frac{y_1 y_2}{1 + y_2}, \frac{y_1}{1 + y_2}\right)\frac{|y_1|}{(1 + y_2)^2} \\ &= \lambda^2 e^{-\lambda y_1}\frac{y_1}{(1 + y_2)^2}, \end{aligned}$$

for $y_1, y_2 \geq 0$. $\qquad\square$

♠ **Exercise 5.6.7.** Show that Y_1 and Y_2 in this example are independent, and find their marginal densities.

♠ **Exercise 5.6.8.** Strictly speaking, the function g in the previous example is not defined when $x_2 = 0$. Can you explain why this is not an important issue here?

♠ **Exercise 5.6.9.** If X and Y have joint density f, show that the density of XY is given by

$$f_{XY}(u) = \int_{-\infty}^{\infty} f(x, u/x)|x|^{-1}dx.$$

To do this, it is wise to first compute the joint density of X and XY.

♠ **Exercise 5.6.10.** Let $X_1 = aY_1 + bY_2$ and $X_2 = cY_1 + dY_2$, where (X_1, X_2) is a continuous random vector with joint density f. Show that (Y_1, Y_2) is also a continuous random vector with density

$$g(y_1, y_2) = |ad - bc|f(ay_1 + by_2, cy_1 + dy_2),$$

if $ad - bc \neq 0$.

There is one more piece of theory associated with functions of random variables. In the discrete context, when X and Y are independent random variables, so are $g(X)$ and $h(Y)$, for any functions g and h, see Theorem 2.2.8. When the random variables are continuous, this should also be the case. In order to *prove* this, we need a very weak condition which we will call *regularity*. Readers who are not interested in these details, can safely assume that all functions in this book and of practical use are regular, and apply the forthcoming Theorem 5.6.13 without problems.

Definition 5.6.11. A function g is said to be *regular* if there exist numbers $\cdots < a_{-1} < a_0 < a_1 < \cdots$, with $a_i \to \infty$ and $a_{-i} \to -\infty$ when $i \to \infty$, so that g is continuous and monotone on each interval (a_i, a_{i+1}).

Example 5.6.12. The function given by $x \to \sin x$ is regular; all polynomial functions are regular. An example of a function which is not regular is $x \to 1_{\mathbb{Q}}(x)$. □

Theorem 5.6.13. *Let $X_1, \ldots, X_n$ be independent continuous random variables, and let $g_1, g_2, \ldots, g_n$ be regular functions. Then $g_1(X_1), g_2(X_2), \ldots, g_n(X_n)$ are independent random variables.*

Proof. Assume for simplicity that $n = 2$. It follows from regularity that for all $x \in \mathbb{R}$, we can write

$$A_1 := \{y : g_1(y) \leq x\} = \bigcup_i A_{1,i}(x)$$

and

$$A_2 := \{y : g_2(y) \leq x\} = \bigcup_i A_{2,i}(x),$$

as unions of pairwise disjoint intervals. Therefore, we can write

$$
\begin{aligned}
P(g_1(X_1) \leq x, g_2(X_2) \leq y) &= \sum_i \sum_j P(X_1 \in A_{1,i}(x), X_2 \in A_{2,j}(y)) \\
&= \sum_i \sum_j P(X_1 \in A_{1,i}(x)) P(X_2 \in A_{2,j}(y)) \\
&= \sum_i P(X_1 \in A_{1,i}(x)) \sum_j P(X_2 \in A_{2,j}(y)) \\
&= P(g_1(X_1) \leq x) P(g_2(X_2) \leq y),
\end{aligned}
$$

proving the theorem.   □

Example 5.6.14. Suppose that X and Y are independent random variables. Then also $\sin X$ and $\cos Y$ are independent, since $x \to \sin x$ and $x \to \cos x$ are regular functions.   □

5.7   Sums of Random Variables

In Exercise 5.6.9, it was shown already that when X and Y have a joint density, the product XY is also a continuous random variable with an explicit density. Is $X+Y$ also a continuous random variable? The answer is yes, and there are various ways to see this.

The first proof uses Theorem 5.6.5. By taking (X, Y) and the function $g(x, y) = (x+y, y)$ (which satisfies all requirements) we see that $(X+Y, Y)$ is a continuous random vector. It then follows from Theorem 5.5.3 that $X+Y$ is also a continuous random variable. This trick can be used for many other functions of X and Y, and shows that all these functions lead to continuous random variables.

In the case of the sum $X + Y$, there is, however, a more direct way to arrive at the same conclusion. It is quite simple to compute the density of $X+Y$ directly. This runs as follows. Writing $Z = X + Y$, and $f(x, y)$ for the joint density of X and Y, we have

$$
\begin{aligned}
P(Z \leq z) &= \iint_{\{(x,y):x+y\leq z\}} f(x, y)\,dx\,dy \\
&= \int_{x=-\infty}^{\infty} \int_{y=-\infty}^{z-x} f(x, y)\,dy\,dx \\
&= \int_{u=-\infty}^{\infty} \int_{v=-\infty}^{z} f(u, v - u)\,dv\,du,
\end{aligned}
$$

by the substitution $u = x$, $v = y + x$. Now interchange the order of the integrals and it follows that

$$
f_{X+Y}(z) = \int_{-\infty}^{\infty} f(x, z - x)\,dx. \tag{5.9}
$$

When X and Y are independent, this formula reduces to

$$f_{X+Y}(z) = \int_{-\infty}^{\infty} f_X(x) f_Y(z-x) dx. \tag{5.10}$$

Example 5.7.1 (The gamma distribution). Suppose that $X_1, \ldots, X_n$ are independent and exponentially distributed with parameter $\lambda > 0$. In this example we show that the density of the sum

$$S = X_1 + \cdots + X_n$$

is given by

$$f_S(x) = \frac{\lambda^n}{(n-1)!} x^{n-1} e^{-\lambda x}. \tag{5.11}$$

A random variable with this distribution is said to have a *gamma* distribution with parameters n and λ,

We proceed by induction. For $n = 1$, the density in (5.11) reduces to the exponential density, and hence the claim is true.

Suppose now that the claim is true for $n - 1$, that is, the sum

$$T = X_1 + \cdots + X_{n-1}$$

has a gamma distribution with parameters n and λ. Since T and X_n are independent, we can use (5.10) to find

$$
\begin{aligned}
f_S(z) &= \int_0^z \frac{\lambda^{n-1}}{(n-2)!} x^{n-2} e^{-\lambda x} \cdot \lambda e^{-\lambda(z-x)} dx \\
&= \frac{\lambda^n}{(n-2)!} e^{-\lambda z} \int_0^z x^{n-2} dx \\
&= \frac{\lambda^n}{(n-1)!} z^{n-1} e^{-\lambda z},
\end{aligned}
$$

proving the induction step. We shall come across this distribution in Chapter 7.

$\square$

5.8   More About the Expectation; Variance

At this point it is convenient to say a few more things about expectations. Recall that we have defined the expectation of a continuous random variable X as

$$E(X) = \int_{-\infty}^{\infty} x f(x) dx,$$

whenever this integral is well defined. We have also seen already that

$$E(aX + b) = aE(X) + b,$$

which we proved by first computing the density of the random variable $aX + b$.

We will now show that in general, $E(X+Y) = E(X)+E(Y)$, for continuous random variables X and Y.

Theorem 5.8.1. *Let X and Y be continuous random variables with finite expectations. Then*

$$E(X + Y) = E(X) + E(Y).$$

Proof. There are various ways to prove this. The route that we follow proceeds via the density f_{X+Y} of the sum $X + Y$, which we obtained in the Section 5.7. We write f for the joint density of X and Y. We then have

$$
\begin{aligned}
E(X + Y) &= \int_{-\infty}^{\infty} z f_{X+Y}(z)\,dz \\
&= \int_{z=-\infty}^{\infty} z \int_{x=-\infty}^{\infty} f(x, z - x)\,dx\,dz \\
&= \int_{y=-\infty}^{\infty} \int_{x=-\infty}^{\infty} (x + y) f(x, y)\,dx\,dy \\
&= \int_{\mathbb{R}} \int_{\mathbb{R}} x f(x, y)\,dx\,dy + \int_{\mathbb{R}} y \int_{\mathbb{R}} f(x, y)\,dx\,dy \\
&= \int_{\mathbb{R}} x \int_{\mathbb{R}} f(x, y)\,dy\,dx + \int_{\mathbb{R}} y \int_{\mathbb{R}} f(x, y)\,dx\,dy \\
&= \int_{\mathbb{R}} x f_X(x)\,dx + \int_{\mathbb{R}} y f_Y(y)\,dy = E(X) + E(Y).
\end{aligned}
$$

$\square$

♠ **Exercise 5.8.2.** Extend this result to all appropriate cases where the expectation of X and/or Y is infinite.

The following Theorem 5.8.4 tells us that in order to compute the expectation of $g(X)$, for a suitable function $g : \mathbb{R} \to \mathbb{R}$, we do not need to compute the density (or mass function) of $g(X)$. In the proof of this theorem, we will need the following lemma, the continuous analogue of Exercise 2.3.8.

Lemma 5.8.3. *Suppose that X is a continuous random variable with density f satisfying $f(x) = 0$ for all $x < 0$. Then*

$$E(X) = \int_0^{\infty} (1 - F_X(x))\,dx,$$

where $F_X(x) = P(X \leq x)$ denotes the distribution function of X, as usual.

Proof.

$$\int_0^\infty (1 - F_X(x))dx = \int_0^\infty \int_x^\infty f(y)dydx$$
$$= \int_0^\infty f(y)\int_0^y dxdy$$
$$= \int_0^\infty yf(y)dy = E(X).$$

$\square$

Theorem 5.8.4. *Let X be a continuous random variable with density f. Let g be such that $g(X)$ is a continuous or discrete random variable. Then*

$$E(g(X)) = \int_{-\infty}^\infty g(x)f(x)dx,$$

whenever this integral is well defined.

Proof. We will only give the proof in the case where $g(X)$ is a continuous random variable. Suppose first that $g(x) \geq 0$ for all x. Then the density $f_{g(X)}(y)$ is equal to 0 for $y < 0$ and we can apply Lemma 5.8.3 to $g(X)$. This leads to

$$E(g(X)) = \int_0^\infty P(g(X) > x)dx = \int_0^\infty \int_{\{y:g(y)>x\}} f(y)dydx$$
$$= \int_{-\infty}^\infty \int_0^{g(y)} dxf(y)dy$$
$$= \int_{-\infty}^\infty g(y)f(y)dy,$$

which is what we wanted to prove, still under the assumption that $g \geq 0$ though. For general g we use a trick that is very useful in general, when we want to extend a result from positive to general functions. We define

$$g^+(x) = \max(g(x),0) \text{ and } g^-(x) = -\min(g(x),0).$$

We say that g^+ is the *positive* part of g and g^- the *negative* part of g. Note that both g^+ and g^- are positive functions, and that

$$g(x) = g^+(x) - g^-(x).$$

Now from Theorem 5.8.1 and the first part of this proof, we have

$$E(g(X)) = E(g^+(X) - g^-(X))$$
$$= E(g^+(X)) - E(g^-(X))$$
$$= \int_{\mathbb{R}} g^+(x)f(x)dx - \int_{\mathbb{R}} g^-(x)f(x)dx$$
$$= \int_{\mathbb{R}} g(x)f(x)dx.$$

$\square$

Example 5.8.5. Suppose that X has an exponential distribution with parameter 1, and $Y = X^2$. We can compute the expection of Y as follows:

$$E(Y) = \int_0^\infty x^2 e^{-x} dx$$
$$= 2.$$

□

♠ **Exercise 5.8.6.** Suppose that (X, Y) is a continuous random vector with joint density $f(x, y)$, and that $g : \mathbb{R}^2 \to \mathbb{R}$ is such that $g(X, Y)$ is a continuous or discrete random variable. Prove that

$$E(g(X, Y)) = \int \int_{\mathbb{R}^2} g(x, y) f(x, y) dx dy.$$

♠ **Exercise 5.8.7.** Show, using the previous exercise, that when X and Y are continuous and independent, then

$$E(XY) = E(X)E(Y).$$

Definition 5.8.8. The *variance* var(X) of a random variable X with finite expectation is defined as

$$\text{var}(X) = E(X - E(X))^2.$$

According to Theorem 5.8.4, we can compute the variance of a random variable X with expectation μ and density f via

$$\text{var}(X) = \int_{-\infty}^\infty (x - \mu)^2 f(x) dx.$$

♠ **Exercise 5.8.9.** Show that var$(X) = E(X^2) - (E(X))^2$. This formula is very useful for computations.

♠ **Exercise 5.8.10.** Show that var$(aX + b) = a^2\text{var}(X)$, using the previous exercise.

Example 5.8.11. To compute the variance of a random variable X with a uniform $(0, 1)$ distribution, we proceed as follows. First we compute $\int_0^1 x^2 f(x) dx = \int_0^1 x^2 dx = \frac{1}{3}$. Hence $E(X^2) = \frac{1}{3}$, and var$(X) = 1/3 - (1/2)^2 = 1/12$. □

Example 5.8.12. To compute the variance of a standard normal distributed random variable X, we compute $\int_{-\infty}^\infty x^2 f_X(x) dx = 1$. This can be done with partial integration. Hence $E(X^2) = 1$, and since $E(X) = 0$, it follows that the variance of X is 1.

If X has a normal distribution with parameters μ and σ^2, we use the fact that

$$Y = \frac{X - \mu}{\sigma}$$

has a standard normal distribution. Since $X = \sigma Y + \mu$, this gives, using Exercise 5.8.10, that var$(X) = \sigma^2\text{var}(Y) = \sigma^2$. □

♠ **Exercise 5.8.13.** Compute the variance of an exponentially distributed random variable.

♠ **Exercise 5.8.14.** Show that when X and Y are independent and have a joint density, then

$$\text{var}(X+Y) = \text{var}(X) + \text{var}(Y).$$

Example 5.8.15 (Bivariate normal distribution). Suppose that (X,Y) has a bivariate normal distribution with density

$$f(x,y) = \frac{1}{2\pi\sqrt{(1-\rho^2)}} e^{-\frac{1}{2(1-\rho^2)}(x^2-2\rho xy+y^2)},$$

where ρ is a parameter taking values strictly between -1 and 1. At this point we can investigate the significance is of the parameter ρ. There is a nice interpretation of ρ, and this can be seen when we compute $E(XY)$. Using the same trick as in the proof that f is a density, we find, using Exercise 5.8.6, that

$$
\begin{aligned}
E(XY) &= \int_{-\infty}^{\infty}\int_{-\infty}^{\infty} xy f(x,y)dxdy \\
&= \int_{\mathbb{R}} y \frac{1}{\sqrt{2\pi}} e^{-\frac{1}{2}y^2} \int_{\mathbb{R}} x \frac{1}{\sqrt{2\pi(1-\rho^2)}} e^{-\frac{1}{2(1-\rho^2)}(x-\rho y)^2} dxdy.
\end{aligned}
$$

The inner integral over x is the expectation of a normal distribution with parameter ρy and $1-\rho^2$ and therefore equal to ρy. Hence we find that

$$E(XY) = \rho \int_{\mathbb{R}} y^2 \frac{1}{\sqrt{2\pi}} e^{-\frac{1}{2}y^2} dy = \rho.$$

Definition 5.8.16. The *covariance* of X and Y is defined as

$$
\begin{aligned}
\text{cov}(X,Y) &= E((X-E(X))(Y-E(Y))) \\
&= E(XY) - E(X)E(Y).
\end{aligned}
$$

We see that in this case, the covariance of X and Y is equal to ρ. Hence in some sense, ρ quantifies the dependence between X and Y. □

♠ **Exercise 5.8.17.** Show that in the above example, X and Y are independent if and only if $\rho = 0$. This means that X and Y are independent if and only if $E(XY) = E(X)E(Y)$. This is quite remarkable: in general independence implies $E(XY) = E(X)E(Y)$ but not the other way around. In this case the two notions are equivalent.

5.9 Random Variables Which are Neither Discrete Nor Continuous

So far we have distinguished between discrete and continuous random variables. The following example shows that there are very natural random variables which are neither discrete nor continuous.

Example 5.9.1. Suppose that we enter a post office, and that we want to describe our waiting time until we are served. Call this waiting time X. What would be a reasonable distribution for X? There are two possibilities: either there is no one in the post office and our waiting time is equal to 0, or there is a certain number of customers before us, and our waiting time is positive. If we have to wait, it is natural to assume that the waiting time is a continuous random variable, for instance with an exponential distribution with parameter $\lambda > 0$.

Now observe that the waiting time X is not discrete, since it can take any value in $\mathbb{R}$. On the other hand, it is not continuous either since there is a positive probability that we do not have to wait at all, that is, $P(X = 0) > 0$. We can view X as a *mixture* of a discrete and a continuous random variable.   □

This example leads to the following formal definition:

Definition 5.9.2. Let Y be a discrete random variable, and Z be a continuous random variable. Furthermore, let U be a random variable with distribution $P(U = 1) = p$ and $P(U = 0) = 1-p$, for some $p \in (0,1)$. Suppose that all random variables are independent. Then the random variable X defined by

$$X = UY + (1 - U)Z \qquad (5.12)$$

is called a *mixture* of Y and Z.

We should think of X as being equal to Y with probability p, and equal to Z with probability $1 - p$. Hence, another way of defining X is as follows:

$$X = \begin{cases} Y & \text{with probability } p, \\ Z & \text{with probability } 1 - p. \end{cases}$$

We can define expectations of mixtures in the obvious way as follows:

Definition 5.9.3. The expectation of the mixture X in (5.12) is defined as

$$E(X) = pE(Y) + (1 - p)E(Z).$$

♠ **Exercise 5.9.4.** Define the variance of the mixture X in (5.12). Motivate your suggestion.

Example 5.9.5. Suppose that the waiting time in a post office is 0 with probability $\frac{1}{4}$, and exponentially distributed (with parameter λ) with probability $\frac{3}{4}$. The expected waiting time is now equal to $\frac{1}{4} \times 0 + \frac{3}{4}\frac{1}{\lambda} = \frac{3}{4\lambda}$. The probability that the waiting time is at most 1 is now equal to $\frac{1}{4} \times 1 + \frac{3}{4} \times P(Z < 1)$, where Z has an exponential distribution with parameter λ.   □

♠ **Exercise 5.9.6.** Finish the computation in this example, and also compute the probability that the waiting time is at least 2.

Mixtures are natural examples of random variables which are neither discrete nor continuous. There is, however, also a class of random variables and vectors which are neither discrete nor continuous, but which are of a different nature, and which cannot be expressed as a combination of discrete and continuous random variables. Rather then try to make a formal definition, we give an example, hoping that this makes the idea clear. The simplest example is in dimension two.

Example 5.9.7. Suppose that X has a uniform distribution on $(-1, 1)$, and consider the vector $(X, |X|)$. Clearly, $(X, |X|)$ is not a discrete random vector, and it is not a mixture of a continuous and discrete random vector either. Nevertheless, we claim that the vector $(X, |X|)$ has no joint density. To see this, we proceed by contradiction.

Suppose, therefore, that it has joint density f. Observe that

$$P((X, |X|) \notin \{(x, y) : y \neq |x|\}) = 0,$$

so that

$$\int\int_{\{(x,y):y\neq|x|\}} f(x, y)dxdy = 0,$$

and therefore

$$\int\int_{\{(x,y):y=|x|\}} f(x, y)dxdy = 1,$$

which is clearly impossible.   □

♠ **Exercise 5.9.8.** Show that for the same X as in Example 5.9.7, the vector (X, X^2) has no joint density.

5.10   Conditional Distributions and Expectations

Suppose that X and Y have a joint density. As in the discrete setting, we would like to talk about the conditional distribution of Y given that X takes the value x, say. However, since $P(X = x) = 0$, we cannot define

$$P(Y \leq y|X = x)$$

as in the discrete setting. This means that the theory as developed in Chapter 2, is not useful here. But clearly, we would like to have a notion of conditional distributions, even in the continuous case. How can we do that? There are, in fact, several possible approaches to this problem.

One can try to define $P(Y \leq y|X = x)$ as the limit of $P(Y \leq y|x \leq X \leq x+\Delta)$, for $\Delta \downarrow 0$. Under certain weak regularity conditions (which I do not specify here), this approach leads to the following computation, assuming that X and Y

have joint density $f(x,y)$:

$$
\begin{aligned}
\lim_{\Delta \downarrow 0} P(Y \leq y | x \leq X \leq x + \Delta) &= \lim_{\Delta \downarrow 0} \frac{P(Y \leq y, x \leq X \leq x + \Delta)}{P(x \leq X \leq x + \Delta)} \\
&= \lim_{\Delta \downarrow 0} \frac{\int_{-\infty}^{y} \int_{x}^{x+\Delta} f(u,v) du\, dv}{\int_{x}^{x+\Delta} f_X(u) du} \\
&\approx \lim_{\Delta \downarrow 0} \frac{\Delta \int_{-\infty}^{y} f(x,v) dv}{\Delta f_X(x)} \\
&= \int_{-\infty}^{y} \frac{f(x,v)}{f_X(x)} dv.
\end{aligned}
$$

Hence this approach leads to the conclusion that we should define

$$
P(Y \leq y | X = x) = \int_{-\infty}^{y} \frac{f(x,v)}{f_X(x)} dv, \tag{5.13}
$$

and hence that the conditional distribution of Y given $X = x$ (as a function of y) has density $f(x,\cdot)/f_X(x)$.

We now explain a different approach which is slightly more general and which is motivated by some aspects of the discrete theory which we discuss first.

In the discrete setting, we have

$$
\begin{aligned}
P(Y \leq y) &= \sum_{x} P(Y \leq y | X = x) P(X = x) \\
&= \sum_{x} F_{Y|X}(y|x) p_X(x).
\end{aligned}
$$

This suggests that in the continuous setting, we could perhaps define the conditional distribution function of Y given $X = x$, denoted by $F_{Y|X}(y|x)$, implicitly via

$$
\int_{-\infty}^{\infty} F_{Y|X}(y|x) f_X(x) dx = P(Y \leq y). \tag{5.14}
$$

This will indeed be our approach. However, relation (5.14) alone is not enough to determine the conditional distribution uniquely. We must demand more relations to essentially guarantee uniqueness. It is instructive to do this in the discrete setting first.

Let X and Y be discrete random variables. The following result was not stated in Chapter 2, but for our current purpose it is very useful.

Theorem 5.10.1. *For discrete random variables X and Y, the conditional distribution functions $F_{Y|X}(\cdot|x)$ satisfy*

$$
\sum_{x \in A} F_{Y|X}(y|x) P(X = x) = P(Y \leq y, X \in A), \tag{5.15}
$$

for any y and A. Moreover, if, for fixed y, there is another function $\tau(x)$ which also, for all A, satisfies

$$\sum_{x\in A} \tau(x)P(X = x) = P(Y \leq y, X \in A),\tag{5.16}$$

then $P(F_{Y|X}(y|X) = \tau(X)) = 1$.

Proof. The proof of the first assertion is similar to the proof of Theorem 2.5.7 and is not repeated here.

The second assertion is proved by contradiction. Suppose $\tau(x)$ exists with $P(F_{Y|X}(y|X) \neq \tau(X)) > 0$. Without loss of generality, we may assume that $P(F_{Y|X}(y|X) > \tau(X)) > 0$. This means that there is an x with $P(X = x) > 0$ and $F_{Y|X}(y|x) > \tau(x)$. Now, taking $A = \{x : F_{Y|X}(y|x) > \tau(x)\}$, we find

$$\sum_{x\in A} F_{Y|X}(y|x)P(X = x) > \sum_{x\in A} \tau(x)P(X = x),$$

which contradicts (5.15) and (5.16).   □

This means that in the discrete setting, we could have defined $F_{Y|X}(y|x)$ as any function $\tau(x)$ which satisfies

$$\sum_{x\in A} \tau(x)P(X = x) = P(Y \leq y, X \in A),\tag{5.17}$$

for all A.

We now return to the continuous setting. Readers not particularly interested in details can skip the following theorem, and read Definition 5.10.3 immediately.

Let X have a density f_X and let Y be any (discrete or continuous) random variable.

Theorem 5.10.2. *Let X have density f_X. Let τ and τ' be two regular functions which satisfy*

$$\int_A \tau(x)f_X(x)dx = \int_A \tau'(x)f_X(x)dx,\tag{5.18}$$

in the sense that if one of the integrals exists, so does the other, and in that case they have the same value. Then

$$P(\tau(X) = \tau'(X)) = 1.$$

Proof. We have

$$\int_A (\tau(x) - \tau'(x))f_X(x)dx = 0.$$

Suppose, again without loss of generality that $P(\tau(X) > \tau'(X)) > 0$. Consider the set $A = \{x : \tau(x) > \tau'(x)\}$. Since τ an τ' are both regular, the integral $\int_A f_X(x)dx$ exists, and we find

$$\int_A (\tau(x) - \tau'(x))f_X(x)dx = 0.$$

Since the integrand is non-negative and piecewise continuous, this means that the integrand must be piecewise equal to 0. This implies that $f_X(x) = 0$ for all $x \in A$, except for at most a countable number of isolated points. This in turn leads to

$$\int_A f_X(x)dx = 0,$$

which contradicts the assumption that $P(\tau(X) > \tau'(X)) > 0$. □

This now enables us to define conditional distributions.

Definition 5.10.3. Let X have density f_X and let Y ba some random variable. A *conditional distribution* indexconditional!distribution of Y given X is a family of functions $F_{Y|X}(\cdot|x)$ with the properties that
(1) for each x, $F_{Y|X}(\cdot|x)$ is the distribution function of a random variable,
(2) for each y, $F_{Y|X}(y|\cdot)$ is regular, as a function of x,
(3) $F_{Y|X}(y|x)$ satisfies, for all $a < b$,

$$\int_a^b F_{Y|X}(y|x)f_X(x)dx = P(Y \leq y, a < X < b). \tag{5.19}$$

Observe that for each fixed y, $F_{Y|X}(y|X)$ is a random number which is with probability one the same for any two choices of the conditional distribution, according to Theorem 5.10.2. Hence there is a certain uniqueness in this definition. Note that this approach is quite general in the sense that we have not required (X, Y) to be a continuous random vector. In Example 5.10.11 below we will give an interesting application of this.

Having done a certain amount of work to formally define conditional distributions, we can now see in a number of examples how this works in practice. The general approach in practice is often to guess the form of the conditional distribution, and after that verify (5.19). Most practical cases will be covered by the following special case in which X and Y have a joint density. It will be understood from now on that all relevant densities are regular.

Example 5.10.4. Suppose that (X, Y) have a joint density f, and let x be such that $f_X(x) > 0$. In this case, the conditional distribution function of Y given $X = x$ is

$$F_{Y|X}(y|x) = \int_{v=-\infty}^y \frac{f(x,v)}{f_X(x)}dv;$$

compare this with (5.13). To check this, we simly plug in this formula into (5.19). This gives

$$\int_a^b \int_{-\infty}^y \frac{f(x,v)}{f_X(x)} dv f_X(x) dx = \int_a^b \int_{-\infty}^y f(x,v) dv dx$$
$$= P(Y \le y, a < X < b),$$

as required. Hence, according to Lemma 5.3.3, the *conditional density* of Y given $X = x$ can be defined by

$$f_{Y|X}(y|x) = \frac{f(x,y)}{f_X(x)},$$

if $f_X(x) > 0$. If $f_X(x) = 0$ we simply put $f_{Y|X}(y|x) = 0$. The *conditional expectation* $E(Y|X = x)$ is now defined as

$$E(Y|X = x) = \int_{-\infty}^\infty y f_{Y|X}(y|x) dy.$$

These formulas should not come as a surprise, as they are the continuous analogue of the discrete formulas.   □

♠ **Exercise 5.10.5.** Make sure that you understand the last remark by writing down the corresponding discrete formulas.

Theorem 5.10.6. *Let X and Y have joint density f. Then we can compute the expection of Y via*

$$E(Y) = \int_{-\infty}^\infty E(Y|X = x) f_X(x) dx.$$

Proof.

$$E(Y) = \int_{x=-\infty}^\infty \int_{y=-\infty}^\infty y f(x,y) dy dx$$
$$= \int_{x=-\infty}^\infty \int_{y=-\infty}^\infty y f_{Y|X}(y|x) f_X(x) dy dx$$
$$= \int_{x=-\infty}^\infty E(Y|X = x) f_X(x) dx.$$

□

Example 5.10.7. Recall the joint density in Example 5.5.5:

$$f_{X,Y}(x,y) = \frac{1}{x},$$

for $0 < y \le x \le 1$, and $f_{X,Y}(x,y) = 0$ otherwise. A simple computation as as in Example 5.10.4 above now shows that the conditional density $f_{Y|X}(y|x)$ is equal to $1/x$, if $0 < y \leq x \leq 1$. This is to say that conditional on X, Y is uniformly distributed on $(0, x)$. We can now calculate the expectation of Y as

$$E(Y) = \int_0^1 E(Y|X = x)f_X(x)dx$$

$$= \int_0^1 \frac{1}{2}xdx = \frac{1}{4}. \qquad \square$$

Example 5.10.8. Let (X, Y) have the standard bivariate normal distribution of Example 5.5.13. A short computation shows that

$$f_{Y|X}(y|x) = \frac{f_{X,Y}(x,y)}{f_X(x)}$$

$$= \frac{1}{\sqrt{2\pi(1-\rho^2)}}e^{-\frac{(y-\rho x)^2}{2(1-\rho^2)}},$$

which we recognise as the density of a normal distribution with parameters ρx and $1 - \rho^2$. It follows that

$$E(Y|X = x) = \rho x. \qquad \square$$

♠ **Exercise 5.10.9.** Let (X, Y) have joint density

$$f(x, y) = \lambda^2 e^{-\lambda y},$$

for $0 \leq x \leq y$, and $f(x, y) = 0$ elsewhere. Find the conditional density and conditional expectation of Y given X.

♠ **Exercise 5.10.10.** Suppose that (X, Y) has a joint density and that X and Y are independent. Show that $f_{X|Y}(x|y) = f_X(x)$.

In the next example, there is no joint density, and yet we will be able to compute conditional probabilities.

Example 5.10.11. Consider the random vector $(X, |X|)$ in Example 5.9.7. The distribution function $F_X(x)$ of X is given by $F_X(x) = \frac{1}{2} + \frac{1}{2}x$, for $-1 < x < 1$. Suppose now that $|X| = y$. What should be the conditional probability that $X \leq x$? Let $x > 0$. Clearly, when $0 < y \leq x$, given that $|X| = y$, the probability that $X \leq x$ must be 1, and when $y > x$, this probability should be $\frac{1}{2}$, by symmetry. We can now verify this by verifying (5.19):

♠ **Exercise 5.10.12.** Show first that $f_{|X|}(x) = 1$ for $0 < x < 1$.

We now can write,

$$\int_0^1 F_{X||X|}(x|y)f_{|X|}(y)dy = \int_0^x dy + \int_x^1 \frac{1}{2}dy$$

$$= x + \frac{1}{2} - \frac{1}{2}x = \frac{1}{2} + \frac{1}{2}x,$$

as required.

♠ **Exercise 5.10.13.** Do a similar computation for the case $x < 0$.

□

   At this point we should halt for a moment and see what we have achieved. We have developed a theory that enables us to compute conditional distributions of Y given X for many random variables X and Y. There are still situations though which cannot be captured satisfactorily with the current machinery. To illustrate this, look at the following worrying but important example.

Example 5.10.14. Consider the exponential random variables X_1 and X_2 in Example 5.6.6. Suppose that we ask for the conditional density of $X_1 + X_2$ given that $X_1 = X_2$. As such, this is a new type of conditioning that does not fit in our framework. But we can perhaps *make* it fit in our framework by looking at

$$Y_2 = X_1/X_2,$$

which was already used in Example 5.6.6. Indeed, conditioning on $X_1 = X_2$ should be the same as conditioning on $Y_2 = 1$. In Example 5.6.6, we showed that $Y_1 = X_1 + X_2$ and Y_2 are independent, and from this it follows that (using Exercise 5.10.10)

$$f_{Y_1|Y_2}(y_1|y_2) = f_{Y_1}(y_1) = \lambda^2 y_1 e^{-\lambda y_1}.$$

This seems to answer our original question. But we could also have translated the original condition $X_1 = X_2$ into the condition that

$$Y_3 := X_1 - X_2 = 0.$$

From this point of view, we are then asked to compute $f_{Y_1|Y_3}(y_1|y_3)$.

♠ **Exercise 5.10.15.** Show that

$$f_{Y_1|Y_3}(y_1|y_3) = \lambda e^{-\lambda(y_1-|y_3|)},$$

for $|y_3| \leq y_1$.

   The required conditional density should therefore be equal to

$$f_{Y_1|Y_3}(y_1|0) = \lambda e^{-\lambda y_1},$$

which is different from our first answer! This seems to be a serious problem. Which of the two answers is correct and why?

   The answers to these questions are not so easy to understand. The problem is that there is no unique correct answer. Both answers are in some sense correct, but they express something different. The key point is that in our current theory of conditioning, we do not condition on a single event anymore. The conditional distribution function $F_{Y|X}(y|x)$ is defined as a *family* of functions, and the definition simply does not make sense for an individual x. Therefore, it is in some sense no longer possible to talk about the conditional probability for an individual x, but only about a conditional probability for all outcomes of the random variable X simultaneously. In our last example, the two random variables Y_2 and Y_3 on which we conditioned were different, and therefore it should, after all, not come as a surprise that the answers turn out to be different. □

5.11 The Law of Large Numbers

In Section 4.1 we proved a number of limit theorems for discrete random variables. It will come as no surprise that Theorem 4.1.1 and Theorem 4.1.4 are true also for continuous random variabes, without any change in the statements. It is not so useful to repeat the proofs here completely. In fact, it is a very good exercise to do this for yourself. In the proof of Theorem 4.1.1 you will need Chebyshev's inequality 2.3.27.

♠ **Exercise 5.11.1.** Give a proof of Chebyshev's inequality for continuous random variables.

♠ **Exercise 5.11.2.** Prove Theorem 4.1.1 and Theorem 4.1.4 for continuous random variables.

Example 5.11.3 (Numerical integration). Let $g : [0,1] \to \mathbb{R}$ be a function with $\int_0^1 g(x)dx < \infty$. We can use the law of large numbers to estimate $\int_0^1 g(x)dx$ numerically as follows.

Let $X_1, X_2, \ldots$ be independent and identically distributed, uniform $(0,1)$ random variables. This is to say that their common density $f_X(x)$ satisfies $f_X(x) = 1$ for $0 < x < 1$. Let $Y_i = g(X_i)$, for $i = 1, 2 \ldots$ We approximate $\int_0^1 g(x)dx$ by

$$I_n = \frac{1}{n} \sum_{i=1}^n Y_i.$$

Now observe that $E(Y_i) = \int_0^1 g(x)f_X(x)dx = \int_0^1 g(x)dx$. It then follows from the general weak law of large numbers that for all $\epsilon > 0$,

$$P\left(\left|I_n - \int_0^1 g(x)dx\right| > \epsilon\right) \to 0,$$

as $n \to \infty$. Hence, for n large enough, I_n is with high probability a good approximation of $\int_0^1 g(x)dx$. □

5.12 Exercises

Exercise 5.12.1. Suppose we choose a completely random point in the interval $(-2, 1)$, and denote the distance to 0 by X. Show that X is a continuous random variable, and compute its density and expectation.

Exercise 5.12.2. Suppose that X is a continuous random variable with density $f(x) = x^3$, for $0 < x < t$, and $f(x) = 0$ elsewhere.
(a) Compute t.
(b) Compute the expectation of X.
(c) Compute $P(X > 1)$.

Exercise 5.12.3. Let X be a continuous random variable with $E(X) = 3/5$ and density $f(x) = a + bx^2$, for $0 < x < 1$, and $f(x) = 0$ elsewhere. Compute a and b.

Exercise 5.12.4. Let X be a continuous random variable with density $f(x) = cx^2$, $x \in (0, 1)$, and $f(x) = 0$ for other values of x.
(a) Compute c.
(b) Compute $E(X)$.
(c) Compute $P(X > \frac{1}{2})$.
(d) Compute the variance of X.

Exercise 5.12.5. Let X be a continuous random variable with density $f(x) = c \sin x$, for $x \in (0, \pi)$, and $f(x) = 0$ elsewhere.
(a) Compute c.
(b) Compute $E(X)$.
(c) Can you think of an example where this distribution arises?

Exercise 5.12.6. Suppose that $b > 0$ and $a < 0$, and let X be a continuous random variable with density

$$f(x) = ax + b,$$

for $x \in (0, -\frac{b}{a})$, and $f(x) = 0$ elsewhere.
(a) Show that this is a density for all a and b which satisfy $b^2 = -2a$.
(b) Compute $E(X)$ for these values of a and b.
(c) Compute $P(X > E(X))$.

Exercise 5.12.7. Let X have density $f_X(x) = cx^2(1 - x)^3$, for $x \in [0, 1]$, and $f_X(x) = 0$ elsewhere.
(a) Compute c.
(b) Compute $P(X > 1/2)$.

Exercise 5.12.8. Let Z be a standard normal random variable. Show that for $x > 0$
(a) $P(Z > x) = P(Z < -x)$;
(b) $P(|Z| > x) = 2P(Z > x)$.

Exercise 5.12.9. If X is an exponential random variable with parameter λ, and $c > 0$, show that cX is exponentially distributed with parameter λ/c.

Exercise 5.12.10. Two people agreed to meet each other on a particular day, between 5 and 6 pm. They arrive (independently) on a uniform time between 5 and 6, and wait for 15 minutes. What is the probability that they meet each other?

Exercise 5.12.11. Let X and Y have joint density $f(x, y) = e^{-x-y}$, for $x, y > 0$. Are X and Y independent? Find the marginal distributions of X and Y and compute their covariance.

Exercise 5.12.12. Let X and Y be independent random variables with an exponential distribution with parameters μ and λ. Let $U = \min\{X, Y\}$ and $V = \max\{X, Y\}$. Finally, let $W = V - U$.
(a) Compute $P(X \leq Y) = P(U = X)$.
(b) Show that U and W are independent.

Exercise 5.12.13. Let X have an exponential distribution with parameter λ. Show that
$$P(X > s + x | X > s) = P(X > x).$$

Explain why this is called the *lack of memory property* of the exponential distribution.

Exercise 5.12.14. The random variable X has a *double-exponential* distribution with parameter $\lambda > 0$ if its density is given by
$$f_X(x) = \frac{1}{2} \lambda e^{-\lambda|x|},$$

for all x. Show that $E(X) = 0$ and $\text{var}(X) = \frac{2}{\lambda^2}$.

Exercise 5.12.15. If X has the double-exponential distribution of the previous exercise, show that $|X|$ has an exponential distribution with parameter λ.

Exercise 5.12.16. Let X be a random variable with finite expectation and variance. Show that
$$\text{var}(X) \leq E(X - a)^2,$$

for all $a \in \mathbb{R}$.

Exercise 5.12.17. When we enter a shop, the number of customers ahead of us has a Poisson distribution with parameter λ. The service time of the customers is exponential with parameter λ, and all random variables involved are independent. Compute the expected waiting time.

Exercise 5.12.18. Let (X, Y) have joint density $f(x, y) = e^{-y}$, for $0 < x < y$, and $f(x, y) = 0$ elsewhere.
(a) Compute the marginal density of Y.
(b) Show that $f_{X|Y}(x, y) = 1/y$, for $0 < x < y$.
(c) First compute $E(X|Y = y)$ and use this to compute $E(X)$.

Exercise 5.12.19. Let (X, Y) have joint density $f(x, y) = xe^{-x(y+1)}$, for $x, y > 0$ and $f(x, y) = 0$ elsewhere.
(a) Compute the marginal distributions of X and Y.
(b) Show that XY has an exponential distribution with parameter 1.
(c) Compute $E(X|Y = y)$.

Exercise 5.12.20. Let (X, Y) have joint density $f(x, y) = xe^{-x-y}$, when $x, y > 0$, and $f(x, y) = 0$ elsewhere. Are X and Y independent?

Exercise 5.12.21. Let X and Y be independent exponentially distributed random variables with parameter 1. Let $Z = Y/X$.
(a) Compute the distribution function of (X, Z).
(b) Are X and Z independent?
(c) Compute $E(Z|X = x)$.

Exercise 5.12.22. Let U and V be independent and uniformly distributed on $(0, 1)$. Let $X = U + V$ and $Y = UV$.
(a) Compute the density of (X, Y), X and Y.
(b) Are X and Y independent?

Exercise 5.12.23. Let X be a $(0, \frac{\pi}{2})$ uniform random variable. Find the density of $Y = \sin X$.

Exercise 5.12.24. Let X and Y have the bivariate normal distribution with density

$$f(x, y) = \frac{1}{2\pi\sqrt{(1 - \rho^2)}} e^{-\frac{1}{2(1-\rho^2)}(x^2 - 2\rho xy + y^2)}.$$

(a) Show that X and $Z = (Y - \rho X)/(1 - \rho^2)^{\frac{1}{2}}$ are independent standard normal random variables.
(b) Deduce from (a) that

$$P(X > 0, Y > 0) = \frac{1}{4} + \frac{1}{2\pi} \arcsin \rho.$$

Exercise 5.12.25. Let X, Y and Z be independent $(0, 1)$ uniform random variables.
(a) Find the joint density of XY and Z^2.
(b) Show that $P(XY < Z^2) = \frac{5}{9}$.

Exercise 5.12.26. Show that the density of the sum of two independent standard normal random variables is given by

$$f(x) = \frac{1}{2\sqrt{\pi}} e^{-\frac{1}{4}x^2}.$$

Exercise 5.12.27. Let X be a continuous random variable with continuous distribution function F. Show that
(a) $F(X)$ is uniformly distributed on $(0, 1)$;
(b) $-\log F(X)$ is exponentially distributed.

Exercise 5.12.28. Let F be a strictly increasing, continuous distribution function, and let U be a uniform $(0, 1)$ distributed random variable. Show that $F^{-1}(U)$ is a random variable with distribution function F.

Exercise 5.12.29. Suppose that X has a Cauchy distribution. Show that $1/X$ also has a Cauchy distribution.

Exercise 5.12.30. Let X and Y be independent with a standard normal distribution. Show that X/Y has a Cauchy distribution.

Exercise 5.12.31. Let (X, Y) have joint density $f(x, y) = \frac{2}{x} e^{-2x}$, for $0 < y < x$, and $f(x, y) = 0$ elsewhere. Compute $\text{cov}(X, Y)$.

Exercise 5.12.32. Compute the conditional distribution of Y given $X = x$ in the previous exercise and compute $E(Y \mid X = x)$.

Exercise 5.12.33. Suppose X has an exponential distribution with parameter $\lambda > 0$. Given $X = x$, we let Y be a uniform point in $(0, x)$. Compute $E(Y)$.

Exercise 5.12.34. Let X and Y be independent and uniformly distributed on $(0, 1)$. Compute the density of $X + Y$.

Exercise 5.12.35. Show that for X and Y independent and standard normal distributed, $\frac{1}{2}(X + Y)$ has a normal distribution with parameters 0 and $\frac{1}{2}$.

Exercise 5.12.36. Let X and Y be independent standard normal random variables, and let $Z = \frac{1}{2}(X^2 + Y^2)$. Compute the density of Z.

Exercise 5.12.37. Let X and Y be independent random variables with the same density $f(x) = a(1 + x^4)^{-1}$. Show that $\arctan(Y/X)$ has a uniform distribution on $(-\pi/2, \pi/2)$.

Exercise 5.12.38. Let X_1, X_2 and X_3 be independent uniform $(0, 1)$ random variables. What is the probability that we can form a triangle with three sticks of length X_1, X_2 and X_3?

Exercise 5.12.39 (Buffon's needle). A plane is ruled by the lines $y = n$ ($n \in \mathbb{Z}$) and a needle of unit length is thrown randomly on to the plane. What is the probability that it intersects a line? In order to answer this question, we have to make a few assumptions. Let Z be the distance from the centre of the needle to the nearest line beneath it, and let Θ be the angle, modulo π, made by the needle and the x-axis.
(a) Explain why it is reasonable to assume that Z is uniform on $(0, 1)$, Θ is uniform om $(0, \pi)$, and that Z and θ are independent.
(b) Show that under the assumptions in (a), the vector (Z, Θ) has joint density

$$f(z, \theta) = \frac{1}{\pi},$$

for all appropriate z and θ.
(c) Show that the needle intersects a line if and only if

$$(Z, \Theta) \in \left\{ (z, \theta) : z \le \frac{1}{2} \sin \theta \text{ or } 1 - z \le \frac{1}{2} \sin \theta \right\}.$$

(d) Show that the probability that the needle intersects a line is equal to $2/\pi$.
(e) Explain how you can use (d) to approximate π numerically. (Think of the law of large numbers.)

Exercise 5.12.40. Let X be uniform on $(0, 1)$. Compute $P(X \le x \mid X^2 = y)$.

Chapter 6

Infinitely Many Repetitions

In the Intermezzo, we found that an infinitely fine operation like choosing a point on a line segment can not be captured with countable sample spaces, and hence we extended our notion of a sample space as to be able to deal with that kind of operations.

However, there is yet another problem, namely how to deal with *infinitely many repetitions* of an experiment. For instance, we might want to talk about an infinite sequence of independent coin flips. In this chapter we look at the problem of how to model and construct an infinite sequence of independent random variables.

Maybe you have the impression that infinitely many coin flips should not be *that* important. Granted, we will never be able to perform infinitely many such flips, and so far we didn't really needed to do this, not even in the laws of large numbers in Section 1.6 and 4.1. In these laws of large numbers, we *first* took n random variables, then computed a probability, and *finally* took the limit for $n \to \infty$. There was no need for infinitely many independent random variables.

Sometimes, however, we want to take the limit *first* and after that compute a probability. So, for instance, we would like to talk about the probability that the average outcomes converge to an expectation. And sometimes it is not clear in advance how many independent random variables with a certain distribution we need. For situations like this, we need to define infinitely many independent random variables on one and the same probability space. In the first two sections of this chapter, we shed some light on this. If you are not particularly interested in these details, it is safe to continue with Section 6.3 with the knowledge that infinitely many independent random variables $X_1, X_2, \ldots$ with any distribution can indeed be defined on some probability space. In the subsequent sections, we discuss some applications, notably strong laws of large numbers, random walk and branching processes.

6.1 Infinitely Many Coin Flips and Random Points in $(0, 1]$

In this section we will show that it is, to some extent, possible to define infinitely many coin flips on a very familiar sample space, namely on $\Omega = (0, 1]$. We start by recalling from Chapter 5 one aspect of the selection of a completely random point from Ω, because we need this for the coin flips: the probability that such a random point is chosen in

$$A = \bigcup_{i=1}^{\infty} (a_i, b_i],$$

where the intervals $(a_i, b_i]$ are pairwise disjoint, was defined as

$$P(A) = \sum_{i=1}^{\infty} (b_i - a_i). \tag{6.1}$$

How does this relate to coin flips? Well, as before, we associate with each $\omega \in \Omega$ its *binary expansion*:

$$\omega = \sum_{n=1}^{\infty} \frac{d_n(\omega)}{2^n} = .d_1(\omega) d_2(\omega) \cdots$$

If ω has more than one such expansion, we take the one with infinitely many 1s. So for instance $1/2$ can be written as $.1000000 \cdots$ but we make the convention to use $1/2 = .01111111 \cdots$ Now consider the sequence

$$(d_1(\omega), d_2(\omega), \cdots).$$

We claim that this sequence behaves, to some extent, as if it resulted from an infinite sequence of flips of a fair coin. To see what we mean by this, note that

$$\{\omega : d_i(\omega) = u_i, i = 1, \ldots, n\} = \left(\sum_{i=1}^{n} \frac{u_i}{2^i}, \sum_{i=1}^{n} \frac{u_i}{2^i} + \frac{1}{2^n} \right], \tag{6.2}$$

where the u_i's take values in 0 and 1. To understand (6.2), one can perhaps first look at those ωs which satisfy $\omega_1 = 0$. The set of ωs that satisfy $\omega_1 = 0$ form the set $(0, \frac{1}{2}]$. Among these ωs, only those in $(\frac{1}{4}, \frac{1}{2}]$ have $\omega_2 = 1$, etcetera.

But now we see that we can assign a probability to the set $\{\omega : d_i(\omega) = u_i, i = 1, \ldots, n\}$, using (6.1) above. Indeed, the length of the interval at the right hand side of (6.2) is 2^{-n}, and we conclude that

$$P(\omega : d_i(\omega) = u_i, i = 1, \ldots, n) = \frac{1}{2^n}.$$

Interpreting $d_i(\omega)$ as the outcome of the ith flip of a coin, we see, indeed, that the probability of a given sequence of outcomes of length n is equal to 2^{-n}, which is what our probabilistic intuition requires.

This means that we have defined, with a very simple sample space, a mathematical model that represents *infinitely* many coin flips. The models for coin flips in earlier chapters were only able to model finitely many coin flips.

Infinitely many events are said to be *independent* if each finite subcollection is independent. In the present context this corresponds to independence of any collection $\{d_i = u_i, i \in I\}$, for any finite index set I. The next exercise shows that indeed, different coin flips are independent.

♠ **Exercise 6.1.1.** Show that

$$P(d_i = u_i, i \in I) = \prod_{i \in I} P(d_i = u_i),$$

for any finite index set I and numbers $u_i \in \{0,1\}$.

We have performed a most remarkable construction. We simultaneously constructed a model for picking a point randomly from $(0,1]$, and a model for infinitely many coin flips. Many interesting events do have probabilities in this simultaneous model.

Example 6.1.2. Consider the event that we see k heads among the first n flips. We know from Example 1.2.10 that the probability of this event should be $\binom{n}{k}2^{-n}$. In the present context we can rederive this result as follows. The event

$$\left\{ \omega : \sum_{i=1}^{n} d_i(\omega) = k \right\}.$$

of seeing k heads among the first n flips is the union of those intervals in (6.2) that have k of the u_i's equal to 1 and $n - k$ equal to 0. There are $\binom{n}{k}$ of such intervals, and each of them has length 2^{-n}, which leads to the required probability. □

Can we assign a probability to *all* reasonable subsets of Ω now? The answer is no. There are very interesting and important subsets which can not be written as such a countable union. As an example, consider the set

$$L = \left\{ \omega : \lim_{n \to \infty} \frac{1}{n} \sum_{i=1}^{n} d_i(\omega) = \frac{1}{2} \right\}. \tag{6.3}$$

This set cannot be expressed as a union of countably many intervals.

♠ **Exercise 6.1.3.** Can you prove this?

In an attempt to assign probabilities to a larger collection of sets, hopefully including L, we make the following definition.

Definition 6.1.4. A subset A of Ω is called *small* if for each $\epsilon > 0$, there exists a finite or countable collection $I_1, I_2, \ldots$ of intervals (possibly overlapping) satisfying

$$A \subseteq \bigcup_k I_k \tag{6.4}$$

and

$$\sum_k P(I_k) < \epsilon. \tag{6.5}$$

It is reasonable to assign probability 0 to small sets, and this is precisely what we do now.

Definition 6.1.5. For any small set A, we define $P(A) = 0$, and $P(A^c) = 1$.

In Section 6.3 below we shall prove that L^c is small, and hence, $P(L) = 1$.

6.2   A More General Approach to Infinitely Many Repetitions

In the last section, we formulated a construction of infinitely many independent coin flips. Beautiful as the construction was, it was also rather limited in its scope. In this section we sketch a very simple construction which allows for infinitely many independent random variables with arbitrary distributions. For this, we need a new sample space.

Consider $\Omega = \mathbb{R}^\infty$, the set whose elements can be written as

$$\omega = (\omega_1, \omega_2, \omega_3, \dots),$$

with $\omega_i \in \mathbb{R}$, for all i. Also let, for all $i = 1, 2, \dots$, f_i be a density on $\mathbb{R}$.

Given this sequence of densities, how can we assign probabilities to suitable subsets of Ω? The idea is that we first only look at so called *finite-dimensional subsets A*. A subset $A \subseteq \Omega$ is called finite-dimensional if we can decide whether or not $\omega \in A$ by only looking at a (nonrandom) finite set of indices. Here is an example.

Example 6.2.1. The set $\{\omega \in \Omega : \omega_2 + \omega_3 \le 5\}$ is a finite-dimensional set, since we need only look at the second and third coordinate to decide whether or not $\omega \in A$. The set $\{\omega \in \Omega : w_i = 0 \text{ for all even indices}\}$ is not finite-dimensional. □

Since any finite-dimensional event depends on only finitely many indices, there is, for any such event, an index n such that A depends only in the indices $1, \dots, n$. Hence, any finite-dimensional event A can, for some n, be written as

$$A = A_n \times \prod_{i=n+1}^{\infty} \mathbb{R},$$

where $A_n \subseteq \mathbb{R}^n$. For this finite-dimensional event, we *define* the probability $P(A)$ of A as

$$P(A) = \int \cdots \int_{A_n} f_1(x_1) \cdots f_n(x_n) dx_1 \cdots dx_n.$$

In this way, we assign a probability to any finite-dimensional subset A.

Example 6.2.2. Let, for all i, f_i be the uniform density on $(0, 1)$, that is, $f_i(x) = 1$ for $0 < x < 1$, and $f_i(x) = 0$ elsewhere. Consider the set $A \subset \Omega$ defined by

$$A = \left\{ \omega \in \Omega : \omega_1 > \frac{1}{2}, \omega_2 < \frac{3}{4} \right\}.$$

Then A is finite-dimensional, and $P(A) = \frac{1}{2} \times \frac{3}{4} = \frac{3}{8}$. □

Example 6.2.3. In the same situation as the previous example, let B be the set

$$B = \{ \omega \in \Omega : \omega_1 + \omega_2 < 1 \}.$$

then B is finite-dimensional, and $P(B) = \frac{1}{2}$. □

We can extend the collection of subsets that receive a probability in two ways, as follows.

1. If $A^1, A^2, \ldots$ are disjoint finite-dimensional sets with union A, we define

$$P(A) = \sum_i P(A^i).$$

2. We define a set to be *small* precisely as in Definition 6.1.4, with intervals replaced by finite-dimensional sets. As before, we assign probability 0 to all small sets, and probability 1 to the complement of a small set.

All subsets that received a probability by this procedure are called *events*, just as before. In the forthcoming sections, all sets under consideration will be events, but we will not mention this explicitly each time.

A *random variable* on Ω is now simply defined as a mapping from Ω to $\mathbb{R}$ such that for all $-\infty \le a \le b \le \infty$, the set

$$\{ \omega \in \Omega : a \le X(\omega) \le b \}$$

is an event. With this construction, we can now define infinitely many independent random variables $X_1, X_2, \ldots$ as follows:

$$X_i(\omega) = \omega_i,$$

for $i = 1, 2, \ldots$ It is clear that X_i is a random variable with density f_i, for all i. It is also clear that

$$P(X_{i_1} \le a_1, \ldots, X_{i_k} \le a_k) = \prod_{j=1}^{k} P(X_{i_j} \le a_j),$$

and hence the X_i's are independent.

♠ **Exercise 6.2.4.** Show that we can also define an infinite sequence of independent random variables with an arbitrary discrete distribution.

6.3   The Strong Law of Large Numbers

In this section we discuss three approaches to what is called the *strong law of large numbers*. The phrase *strong* refers to the fact that in these laws, the limit is *inside* the probability, as opposed to the laws of large numbers that we have seen so far. The first strong law goes back to Theorem 1.6.1, and deals with the situation in which the random variables can only take the values 0 and 1 with equal probability.

Consider independent random variables $X_1, X_2, \ldots$, all with the same distribution given by $P(X_i = 1) = P(X_i = 0) = \frac{1}{2}$, for all i. In (1.5) it was shown (in the current notation) that

$$P\left(\left|\frac{1}{n}\sum_{i=1}^{n}X_i - \frac{1}{2}\right| > \epsilon\right) \leq 2e^{-\epsilon^2 n}, \tag{6.6}$$

where the factor 2 comes from the fact that we need to consider deviations above and below of $\frac{1}{2}$. We use this formula with a surprising trick: we take ϵ depending on n, namely $\epsilon = n^{-1/4}$. The reason for this choice becomes clear in a moment. We obtain

$$P\left(\left|\frac{1}{n}\sum_{i=1}^{n}X_i - \frac{1}{2}\right| > n^{-1/4}\right) \leq 2e^{-\sqrt{n}}. \tag{6.7}$$

Define

$$F = \left\{\lim_{n\to\infty}\frac{1}{n}\sum_{i=1}^{n}X_i = \frac{1}{2}\right\}.$$

We want to show that F^c is small, in the sense of Section 6.2. Denote the event in (6.7) by A_n. A little thought reveals that

$$F^c \subseteq \bigcup_{n=m}^{\infty} A_n, \tag{6.8}$$

for all m. Since $P(A_n) \leq 2e^{-\sqrt{n}}$, the sum $\sum_n P(A_n)$ is finite (this is the reason for our choice of ϵ above), and hence $\sum_{n=m}^{\infty} P(A_n)$ can be made arbitrary small by taking m large. This implies by (6.8) that F^c is small, and hence that $P(F) = 1$. We have proved the following result.

Theorem 6.3.1 (First strong law of large numbers). *Let* $X_1, X_2, \ldots$, *be independent random variables, all with the same distribution given by* $P(X_i = 1) = P(X_i = 0) = \frac{1}{2}$, *for all* i. *Then*

$$P\left(\lim_{n\to\infty}\frac{1}{n}\sum_{i=1}^{n}X_i = \frac{1}{2}\right) = 1.$$

With some extra work we can generalise Theorem 6.3.1 considerably, as follows:

Theorem 6.3.2 (Second strong law of large numbers). *Let $X_1, X_2, \ldots$ be independent and identically distributed random variables, such that $E(X_1^4) < \infty$. We write $E(X_1) = \mu$. Then*

$$P\left(\lim_{n \to \infty} \frac{1}{n} \sum_{i=1}^{n} X_i = \mu \right) = 1. \tag{6.9}$$

Proof. We first assume that $\mu = 0$, and we denote the event in (6.9) by M. With Markov's inequality, we can show, writing $S_n = X_1 + \cdots + X_n$, that

$$P\left(\left| \frac{S_n}{n} \right| > \epsilon \right) \leq \frac{E(S_n^4)}{n^4 \epsilon^4}.$$

We now write

$$
\begin{aligned}
E(S_n^4) &= \sum_{i=1}^{n} \sum_{j=1}^{n} \sum_{k=1}^{n} \sum_{l=1}^{n} E(X_i X_j X_k X_l) \\
&= \sum_{i=1}^{n} E(X_i^4) + \sum_{i,j=1, i \neq j}^{n} E(X_i^2) E(X_j^2) \\
&= n E(X_1^4) + 3n(n-1) E(X_1^2) E(X_2^2) \quad \text{(see explanation below)} \\
&\leq cn^2,
\end{aligned}
$$

for some positive constant c. To understand the middle two lines, note that if a certain X_i appears in isolation in the four-fold product, then the corresponding expectation is 0. There are n ways to get a term of the form X_i^4 and $\binom{n}{2} \times 6 = 3n(n-1)$ ways to obtain a term of the form $X_i^2 X_j^2$ for $i \neq j$. All other terms are zero since they contain at least one X_i in isolation.

♠ **Exercise 6.3.3.** Check these last combinatorial statements.

It now follows that

$$P\left(\left| \frac{S_n}{n} \right| > \epsilon \right) \leq \frac{c}{n^2 \epsilon^4}. \tag{6.10}$$

So far, ϵ was independent of n. But now we choose $\epsilon_n = n^{-1/8}$. The reason for this somewhat strange choice is the fact that the series $\sum_n \epsilon_n^{-4} n^{-2}$ converges. Hence, writing A_n for the event in (6.10) we now have that

$$\sum_{n=1}^{\infty} P(A_n) < \infty.$$

If ω is *not* in A_n for all n larger than some m, then $|n^{-1}s_n(\omega)| < \epsilon_n$ for all $n > m$, and since $\epsilon_n \to 0$ as $n \to \infty$, this implies that $\omega \in M$. We conclude that

$$\bigcap_{n=m}^{\infty} A_n^c \subset M,$$

for all m, which is the same as

$$M^c \subset \bigcup_{n=m}^{\infty} A_n.$$

The A_n's form a collection of disjoint intervals, and since $\sum_n P(A_n) < \infty$, we can, for any $\epsilon > 0$, choose m so large that $\sum_{n=m}^{\infty} P(A_n) < \epsilon$. This means that M^c is small.

Note that we have assumed that $\mu = 0$ so far. To prove the general statement, assume that $E(X_1) = \mu$, and define $Y_i = X_i - \mu$, for $i = 1, 2, \ldots$ Then $E(Y_i) = 0$, and according to what we have already proved, it follows that

$$\frac{1}{n} \sum_{i=1}^{n} Y_i \to 0,$$

with probability 1, as $n \to \infty$. This means that

$$\frac{1}{n} \sum_{i=1}^{n} X_i - \mu \to 0,$$

and we have proved the general result.   □

We can also approach the strong law via the set-up of Section 6.1. Recall the set L which was defined as

$$L = \left\{ \omega : \lim_{n \to \infty} \frac{1}{n} \sum_{i=1}^{n} d_i(\omega) = \frac{1}{2} \right\}. \tag{6.11}$$

Theorem 6.3.4 (Third strong law of large numbers). *For L as in (6.11), we have that L^c is a small set, and hence*

$$P(L) = 1.$$

In fact, the proof is very similar to the proof of Theorem 6.3.2. It is instructive, however, to see how a proof in the setup of Section 6.1 proceeds, and we therefore give some details. We first need a preliminary lemma.

Lemma 6.3.5. *Let $0 = x_0 < x_1 < \cdots < x_{k-1} < x_k = 1$, and let $f : (0, 1] \to \mathbb{R}$ be such that f is constant between consecutive x_i's: $f(\omega) = c_j$, for all $\omega \in (x_{j-1}, x_j]$. Then, for $\alpha > 0$, the set $\{\omega : f(\omega) \geq \alpha\}$ is a finite union of intervals and*

$$P(\omega : f(\omega) \geq \alpha) \leq \frac{1}{\alpha} \int_0^1 f(\omega) d\omega.$$

Proof. The set in question is simply the union of those intervals $(x_{j-1}, x_j]$ for which $c_j \geq \alpha$. Hence

$$
\begin{aligned}
\alpha P(\omega : f(\omega) \geq \alpha) &= \alpha \sum_{j:c_j \geq \alpha} (x_j - x_{j-1}) \\
&\leq \sum_{j:c_j \geq \alpha} c_j (x_i - x_{j-1}) \\
&\leq \sum_j c_j (x_i - x_{j-1}) \\
&= \int_0^1 f(\omega) d\omega.
\end{aligned}
$$

$\square$

♠ **Exercise 6.3.6.** Do you the similarity between this lemma and Markov's inequality?

Proof of Theorem 6.3.4. It is convenient to change the digits from 0s and 1s to $+1$s and -1s. Hence we define $r_n(\omega) = 2d_n(\omega) - 1$, that is, if $d_n(\omega) = 1$, then $r_n(\omega) = 1$, and if $d_n(\omega) = 0$, then $r_n(\omega) = -1$. Now consider the partial sums

$$
s_n(\omega) = \sum_{i=1}^n r_i(\omega).
$$

Clearly, L and

$$
N = \left\{ \omega : \lim_{n \to \infty} \frac{1}{n} s_n(\omega) = 0 \right\}
$$

define the same set, and it suffices to show that N^c is small. Applying Lemma 6.3.5 to $f(\omega) = s_n^4(\omega)$ and $\alpha = n^4 \epsilon^4$, we obtain

$$
P(\omega : |s_n(\omega)| \geq n\epsilon) \leq \frac{1}{n^4 \epsilon^4} \int_0^1 s_n^4(\omega) d\omega. \tag{6.12}
$$

Clearly, we have

$$
s_n^4(\omega) = \sum r_i(\omega) r_j(\omega) r_k(\omega) r_l(\omega),
$$

where all four indices range from 1 to n. The terms inside the sum can be of various forms.

If the four indices are all the same, the outcome is clearly 1, and the same is true if they are pairwise the same, that is, of the form $r_i^2(\omega) r_j^2(\omega)$ for $i \neq j$. There are n occurrences of the first type, and $3n(n-1)$ of the second: indeed, there are n choices for the first index i, three ways to match it with j, k or l, and then $n-1$ choices for the value common to the remaining two indices.

If the product is of the form $r_k^2(\omega) r_j(\omega) r_i(\omega)$ for different i, j and k, then this reduces to $r_j(\omega) r_i(\omega)$. Assume without loss of generality that $i < j$. Now observe that on a dyadic interval of size 2^{-j+1}, r_i is constant and r_j has value -1 on the left, and $+1$ on the right. The product $r_i r_j$ therefore integrates to 0 over each of the dyadic intervals of size 2^{-j+1}, and we find that

$$\int_0^1 r_k^2(\omega) r_j(\omega) r_i(\omega) d\omega = 0,$$

if k, i and j are all different.

Since $r_i^3(\omega) r_j(\omega) = r_i(\omega) r_j(\omega)$, also

$$\int_0^1 r_i^3(\omega) r_j(\omega) d\omega = 0,$$

if $i \neq j$.

Finally, a similar argument shows that

$$\int_0^1 r_i(\omega) r_j(\omega) r_k(\omega) r_l(\omega) d\omega = 0.$$

♠ **Exercise 6.3.7.** Prove this last statement.

Putting everything together now gives

$$\int_0^1 s_n^4(\omega) d\omega = \sum_{i=1}^n \int_0^1 r_i^4(\omega) d\omega + \sum_{i,j=1, i \neq j}^n \int_0^1 r_i^2(\omega) r_j^2(\omega) d\omega$$
$$= n + 3n(n-1) \leq 3n^2.$$

It now follows from (6.12) that

$$P\left(\omega : \left|\frac{s_n(\omega)}{n}\right| \geq \epsilon\right) \leq \frac{3}{n^2 \epsilon^4}.$$

The proof is now finished exactly as the proof of Theorem 6.3.2   □

♠ **Exercise 6.3.8.** Make sure that you agree with this last statement.

6.4 Random Walk Revisited

In Chapter 3 we discussed random walk , defined as

$$S_n = \sum_{i=1}^n X_i,$$

where $X_1, \ldots, X_n$ are independent random variables, all with the same distribution given by $P(X_1 = 1) = P(X_1 = -1) = \frac{1}{2}$. In the current context, we can investigate the event that the random walk returns to its starting point 0, that is, we can study the event

$$A = \{S_n \neq 0 \text{ for all } n > 0\}.$$

The following theorem tells us that A has probability 0, in other words, the probability that the random walk returns to 0 is equal to 1. This is called *recurrence* of the random walk.

Theorem 6.4.1 (Recurrence of the random walk). *We have* $P(A) = 0$.

Proof. The proof is very easy, given what we have already done in Chapter 3. Indeed, in the proof of Lemma 3.1.14 we showed that

$$P(S_1 S_2 \cdots S_{2m} \neq 0) = \binom{2m}{m} \left(\frac{1}{2}\right)^{2m}, \tag{6.13}$$

and in the proof of Theorem 3.2.3 we showed that the right hand side of (6.13) tends to 0 as $m \to \infty$. Since

$$A \subseteq \{S_1 S_2 \cdots S_{2m} \neq 0\}$$

for all m, this means that A is small, and therefore has probability 0. $\square$

6.5 Branching Processes

In this section we discuss a few aspects of branching processes . A branching process is a simple model for reproduction and can be described as follows.

Suppose that a population evolves in generations, and denote by Z_n the number of members of the nth generation. Each member of the nth generation gives birth to a random number of children, which will be members of the $(n+1)$th generation. We make two basic assumptions about the number of children:

1. The number of children of different members of the population are independent of each other.

2. The number of children of different members of the population all have the same distribution.

Assuming that $Z_0 = 1$, we are interested in the random variables $Z_1, Z_2, Z_3, \ldots$ The description of the process can be formulated as follows:

$$Z_{n+1} = X_1 + X_2 + \cdots + X_{Z_n}, \tag{6.14}$$

where the X_i's are independent random variables representing the number of children of a particular member of the population. Indeed, each member of the nth generation begets a random number of children, and therefore the number of members of the $(n+1)$th generation is given as in (6.14). Note that the number of X_i's in (6.14) is random. We call the common distribution of the X_i's the *offspring distribution*.

It turns out that a branching process can be well-studied with the help of generating functions, which were defined in Section 2.6. In the current situation we have to deal with a sum of independent random variables of random length. To this end, we state the following lemma.

Lemma 6.5.1. *Let $X_1, X_2, \ldots$ be a sequence of independent identically distributed random variables taking values in $\mathbb{N}$ and with common generating function G_X. Let N be a random variable, independent of the X_i's, also taking values in $\mathbb{N}$, with generating function G_N. Then the sum*

$$S = X_1 + X_2 + \cdots + X_N$$

has generating function given by

$$G_S(s) = G_N(G_X(s)).$$

Proof. We write

$$
\begin{aligned}
G_S(s) &= E(s^S) \\
&= \sum_{n=0}^{\infty} E(s^S | N = n) P(N = n) \\
&= \sum_{n=0}^{\infty} E(s^{X_1 + X_2 + \cdots + X_n}) P(N = n) \\
&= \sum_{n=0}^{\infty} E(s^{X_1}) \cdots E(s^{X_n}) P(N = n) \\
&= \sum_{n=0}^{\infty} (G_X(s))^n P(N = n) = G_N(G_X(s)).
\end{aligned}
$$

$\square$

Writing G_n for the generating function of Z_n, Lemma 6.5.1 and (6.14) together imply that

$$G_{n+1}(s) = G_n(G_1(s)),$$

and iteration of this formula implies that

$$G_n(s) = G_1(G_1(\cdots(G_1(s))\cdots)), \qquad (6.15)$$

the n-fold iteration of G_1. Note that G_1 is just the generating function of an individual X_i, and we write $G = G_1$ from now on. In principle, (6.15) tells us everything about Z_n. For instance, we can now prove the following:

Theorem 6.5.2. *If $E(X_i) = \mu$ and $\mathrm{var}(X_i) = \sigma^2$, then*

$$E(Z_n) = \mu^n,$$

and

$$\mathrm{var}(Z_n) = \begin{cases} n\sigma^2 & \text{if } \mu = 1 \\ \sigma^2(\mu^n - 1)\mu^{n-1}(\mu - 1)^{-1} & \text{if } \mu \neq 1. \end{cases}$$

Proof. Differentiate $G_n(s) = G(G_{n-1}(s))$ at $s = 1$, and use Theorem 2.6.7(a) to find

$$E(Z_n) = \mu E(Z_{n-1}).$$

Now iterate this formula to obtain the first result in the current theorem. Differentiate the same formula twice to obtain

$$G_n''(1) = G''(1)(G_{n-1}'(1))^2 + G'(1)G_{n-1}''(1),$$

and use Theorem 2.6.7(b) to obtain the second result. □

♠ **Exercise 6.5.3.** Provide all details of this last proof.

Hence, the expected number of members in a branching process grows or decays exponentially fast. If the expected numebr of children is larger than 1, the expection grows to infinity, if it is smaller, it decays to zero. This fact should make us curious as to whether it is possible that the branching process survives forever. To this end we first prove:

Theorem 6.5.4. *The probability η that $Z_n = 0$ for some n is equal to the smallest non-negative root of the equation $G(s) = s$.*

Example 6.5.5. Here is an example of Theorem 6.5.4 in action. Consider a branching process with offspring distribution given by $P(X = 0) = \frac{1}{8}$, $P(X = 1) = \frac{1}{2}$ and $P(X = 2) = \frac{3}{8}$. The generating function G is now given by

$$G(s) = \frac{3}{8}s^2 + \frac{1}{2}s + \frac{1}{8}.$$

Solving $G(s) = s$ gives $s = \frac{1}{3}$ and $s = 1$. The smallest non-negative solution is $s = \frac{1}{3}$, and therefore this process survives forever with probability $\frac{2}{3}$. □

Proof of Theorem 6.5.4. The probability η of ultimate extinction can be approximated by $\eta_n = P(Z_n = 0)$. Indeed, it is not hard to see that $\eta_n \to \eta$ as $n \to \infty$. We now write

$$\eta_n = P(Z_n = 0) = G_n(0) = G(G_{n-1}(0)) = G(\eta_{n-1}).$$

Now let $n \to \infty$ and use the fact that G is continuous to obtain

$$\eta = G(\eta).$$

This tells us that η is indeed a root of $G(s) = s$, but the claim is that it is the *smallest* non-negative root. To verify this, suppose that e is any non-negative root of the equation $G(s) = s$. Since G is non-decreasing on $[0, 1]$ we have

$$\eta_1 = G(0) \le G(e) = e,$$

and

$$\eta_2 = G(\eta_1) \le G(e) = e,$$

and so on, giving that $\eta_n \le e$ for all n and hence $\eta \le e$. $\square$

The next result tells us that survival of a branching process is only possible when $\mu > 1$, ignoring the trivial case where $P(X = 1) = 1$, in which case the process trivially survives with probability one.

Theorem 6.5.6. *When $\mu \le 1$, the branching process does not survive with probability one, except for the trivial case where $P(X = 1) = 1$. When $\mu > 1$, the process survives forever with positive probability.*

Proof. According to Theorem 6.5.4, we need to look at the smallest non-negative root of the equation $G(s) = s$.

Suppose first that $\mu > 1$. Since

$$G'(1) = \mu,$$

we have that $G'(1) > 1$. Since $G(1) = 1$, this means that there is some $s' < 1$ for which $G(s') < s'$. Since $G(0) \ge 0$ and since G is continuous, there must be some point s'' between 0 and s' with $G(s'') = s''$, which implies that the smallest non-negative solution of $G(s) = s$ is strictly smaller than 1. Hence the process survives forever with positive probability.

Next, consider the case in which $\mu \le 1$. Note that

$$G'(s) = \sum_{n=1}^{\infty} ns^{n-1}P(X = n) > 0,$$

and

$$G''(s) = \sum_{n=2}^{\infty} n(n-1)s^{n-2}P(X = n) > 0,$$

where the strict inequalities come from the fact that we have excluded the case $P(X = 1) = 1$. This implies that G is strictly increasing and strictly convex. Hence if $G'(1) < 1$, then $G'(s) < 1$ for all $s \in [0, 1]$ and then it is easy to see that $G(s) > s$ for all $s < 1$, and therefore the smallest non-negative solution of $G(s) = s$ is $s = 1$, proving the result. $\square$

6.6 Exercises

Exercise 6.6.1. From $[0,1]$ remove the middle third $(\frac{1}{3}, \frac{2}{3})$. From the remainder, a union of two interval, remove the two open middle thirds $(\frac{1}{9}, \frac{2}{9})$ and $(\frac{7}{9}, \frac{8}{9})$. Continue in the obvious way; what remains when this process is repeated infinitely often is called the *Cantor set*.
(a) Show that the Cantor set consists of exactly those points whose ternary expansion (that is, to the base 3) contains no 1s.
(b) Show that the Cantor set is small in the sense of Definition 6.1.4.

Exercise 6.6.2. Show that $\int_0^1 s_n(\omega)d\omega = 0$.

Exercise 6.6.3. Show that $\int_0^1 s_n^2(\omega)d\omega = n$.

Exercise 6.6.4. Show that any countable subset of $[0,1]$ is small in the sense of Definition 6.1.4.

Exercise 6.6.5. Consider a branching process with a geometric offspring distribution $P(X = k) = (1-p)p^k$, for $k = 0, 1, 2, \ldots$ Show that ultimate extinction is certain if $p \leq \frac{1}{2}$ and that the probability of extinction is $(1-p)/p$ if $p > \frac{1}{2}$.

Exercise 6.6.6. Show that for a branching process (Z_n) with expected offspring μ, we have

$$E(Z_m Z_n) = \mu^{n-m} E(Z_m^2),$$

for $m \leq n$.

Chapter 7

The Poisson Process

In this chapter we discuss a probabilistic model which can be used to describe the occurrences of unpredictable events, which do exhibit a certain amount of *statistical regularity*. Examples to keep in mind are the moments at which telephone calls are received in a call centre, the moments at which customers enter a particular shop, or the moments at which California is hit by an earthquake. We refer to an occurrence of such an unpredictable event simply as an *occurrence*.

7.1 Building a Model

When we try to model the occurrences of the various processes described above, then there are a number of characteristics that we may want to build in. To name a few:

1. There is a certain amount of regularity in the processes described above. Although individual earthquakes are impossible to predict and certainly do not occur in a strictly regular pattern, there is perhaps some statistical regularity in the sense that when we observe the earthquakes during 10 years, say, without knowing the absolute time frame, then we have no way to decide whether we observe the time period 1910-1920 or 1990-2000. In probabilistic terms, the process is *stationary* in time. In other words, the course of time should not change the probabilistic properties of the process. (Of course, in the case of a shop this can only be realistic as long as the shop is open, and even then one can ask whether there will typically be more customers around closing time than around 3 p.m., say. More about this in Exercise 7.5.8)

2. The fact that there is an occurrence at a particular time, says nothing about the probability of an occurrence at, or around, a later or earlier time. In other words, there seems to be some kind of *independence* with respect to various occurrences.

3. The next occurrence can not be predicted form current and past information. In other words, the process of occurrences seems to have *no memory*. The fact that something happened in the past has no effect on the probabilities for future occurrences.

4. There is *no accumulation* of occurrences at any time. In other words, in each finite time interval, there are only finitely many occurrences.

When we try to build a mathematical model for this kind of processes, we should keep these characteristics in mind. In fact, the characteristics pave the way to the appropriate model, as we will demonstrate now. It is quite important that you see how this works. Doing computations in a given model is one thing, but to make an appropriate model is obviously of the highest importance. If the model is not appropriate, then any computation in it has very little, if any, value.

So how can we build a model in the context of the characteristics described above? There are, in fact, a number of ways and we think that it is very instructive to travel them all.

Approach 1 (via waiting times). Perhaps the most natural thing to do is to concentrate on the waiting times. Point (1) above suggests that the waiting time distribution should be the same at all times: the waiting time between the 6th and 7th event should have the same distribution as the waiting time between the 12th and 13th.

Point (3) above suggests that the process should have no memory. We have come across a continuous distribution with a certain lack of memory property. Indeed, in Exercise 5.12.13 we showed that the exponential distribution has no memory in the sense that when X has such an exponential distribution, then

$$P(X > s + t | X > t) = P(X > s).$$

If we think of X as the waiting time between successive occurrences, than this formula expresses the idea that the fact that we have waited already t time units does not change the probability that we have to wait another s time units. This property makes the exponential distribution a serious candidate for the waiting times between successive occurrences.

The candidacy of the exponential distribution becomes even better motivated when we look back at Example 5.1.6. In this example we showed that the exponential distribution is a very natural candidate to model the waiting time for the next occurrence.

Hence we might be inclined to define the following model. We consider independent random variables $X_1, X_2, \ldots$, which are exponentially distributed with parameter λ, that is, they have density f given by

$$f(x) = \lambda e^{-\lambda x},$$

for $x > 0$, and $f(x) = 0$ for $x < 0$. The first occurrence is at time X_1, the second at time $X_1 + X_2$, et cetera. In general, we can define

$$S_n = \sum_{i=1}^{n} X_i,$$

and then S_n should be thought of as the moment of the nth occurrence.

In principle, this defines the model completely. It is possible to do computations now, like computing the probability that there is no occurrence between time 5 and time 10; we come back to this. For now, it is perhaps good to look at the parameter λ. Since $E(X_i) = \lambda^{-1}$, we see that a high λ corresponds to a small average waiting time. Therefore, it makes sense to call λ the *intensity* of the process.

Approach 2 (via the number of occurrences). Another approach becomes apparent when we look back at Example 1.5.13. In that example, we undertook the enterprise to model the arrival of customers in a shop between time $t = 0$ and $t = 1$, not by focussing on the waiting times, but by concentrating on the number of customers. The waiting time should be a continuous random variable, but the number of customers is obviously discrete. Using a discrete approximation with the binomial distribution, we showed that it is reasonable to assume that the probability of having k customers in the shop between time 0 and time 1 is equal to

$$P(N = k) = e^{-\lambda} \frac{\lambda^k}{k!},$$

for $k = 0, 1, \ldots$, which we recognize as the probability mass function of a Poisson distribution with parameter λ.

How can we extend this to longer time intervals? Well, using the idea of independence, mentioned in point (2) above, we might say that the numbers of customers in two disjoint time intervals should be independent. So the number of customers between $t = 1$ and $t = 2$ should be independent of the number of customers between $t = 0$ and $t = 1$, and have the same distribution. As a result, the *total* number of customers in the time interval between $t = 0$ and $t = 2$ can be seen as the sum of two independent Poisson distributed random variables with parameter λ, and we know from Example 2.4.13 that this yields a Poisson distribution with parameter 2λ. This leads to the idea that the number of customers in a time interval of length L should have a Poisson distribution with parameter λL.

This approach does not tell us immediately *when* the customers arrive, it only tells us that in a given time interval the number of customers should have a Poisson distribution, with a parameter which is proportional to the length of the interval. Since we want our process to be stationary, it is reasonable to distribute all customers over the interval in a completely arbitrary way. The model then amounts to the following two-step procedure for defining a process on any time interval of length L:

1. Draw a random number N from a Poisson distribution with parameter λL. This number represents the total number of occurrences in the time interval of length L under consideration.

2. Each of the N customers is given a position, which we choose independently and uniformly distributed over the time interval.

It is not a priori clear that this approach leads to the same model as in Approach 1. We shall see shortly that they are, in fact, equivalent. At this point, we only note the following connection: the probability that there is no occurrence before time t is, according to the current approach, equal to $e^{-\lambda t}$. Now note that this is equivalent to saying that the waiting time for the first occurrence has an exponential distribution with parameter λ, in full agreement with Approach 1.

Approach 3 (via differential equations). Still, there are other ways to approach the problem, without reference to earlier exercises or examples.

We can ask about the probability that at time h (think of h as very small) exactly one occurrence has taken place, or about the probability that by time h, no occurrence, or more than one occurrence has taken place. A natural way to do this proceeds as follows.

Let $p_i(h)$ be the probability that at time h, exactly i occurrences have taken place. Let us assume that $p_1(h)$ is differentiable in $h = 0$, with derivative $\lambda > 0$, that is,

$$p_1'(0) = \lambda. \tag{7.1}$$

The requirement that we do not want accumulation of occurrences can be formulated in terms of $p_{\geq 2}(h) = \sum_{i=2}^{\infty} p_i(h)$ by the requirement that

$$p_{\geq 2}'(0) = 0. \tag{7.2}$$

Since

$$p_0(h) + p_1(h) + p_{\geq 2}(h) = 1,$$

we see, using (7.1) and (7.2), that $p_0(h)$ is differentiable in $h = 0$ with derivative $-\lambda$, that is,

$$p_0'(0) = -\lambda. \tag{7.3}$$

Assuming (7.1), (7.2) and (7.3) we now argue as follows.

$$
\begin{aligned}
p_0(t+h) &= P(\text{no occurrence before } t + h) \\
&= P(\text{no occurrence before } t)P(\text{no occurrence between } t \text{ and } t + h),
\end{aligned}
$$

since we want the number of occurrences in disjoint time intervals to be independent. By stationarity, the probability of having no occurrences between times t and $t + h$ should be the same as $p_0(h)$, and the probability of having no occurrences before time t is by definition equal to $p_0(t)$. Hence we find

$$p_0(t+h) = p_0(t)p_0(h),$$

and a little algebra shows that this leads to

$$\frac{p_0(t+h) - p_0(t)}{h} = p_0(t)\frac{p_0(h) - 1}{h}$$

$$= p_0(t)\frac{p_0(h) - p_0(0)}{h}.$$

Taking the limit for $h \to 0$ and using (7.3) now leads to

$$p_0'(t) = -\lambda p_0(t),$$

and this differential equation is easily solved (using the boundary condition $p_0(0) = 1$), giving

$$p_0(t) = e^{-\lambda t}.$$

Now this is quite interesting, since this expression tells us that the probability that there is no occurrence before time t is equal to $e^{-\lambda t}$. But this means that the probability that the *waiting time* for the first occurrence has an exponential distribution with parameter λ, again in full agreement with Approach 1 and Approach 2.

We see that so far, the various approaches seem to be compatible. They seem to direct us towards a model in which waiting times are continuous with an exponential distribution, and where the total number of occurrences in a given time interval can be described with a Poisson distribution. Is there a first choice? Is one approach better than the other? Hardly, at least at this point (but see Exercise 7.5.9). When we choose a particular approach (for instance Approach 1) then the fact that waiting times are exponentially distributed is true by definition, whereas the fact that the total number of occurrences by time t has a Poisson distribution, has to be proved. If we choose Approach 2, then this last fact becomes true by definition, and we have to prove the statements about the waiting times. Hence the choice of the approach is to a large extent arbitrary.

In this chapter, we choose the first approach, perhaps because this is the easiest conceptually. Hence, we make the following definition.

Definition 7.1.1. Let $X_1, X_2, \ldots$ be independent and indentically distributed random variables with an exponential distribution with parameter $\lambda > 0$. Let $S_0 = 0$, and for $n = 1, 2, \ldots$, let

$$S_n = \sum_{i=1}^{n} X_i.$$

We call X_i the ith *inter-arrival time*, and S_n the moment of the nth occurrence. We define, for all $t \geq 0$, $N(t)$ as the number of occurrences up to and including time t, that is, we define

1. $N(0) = 0$,

2. For $t > 0$, $N(t) = \max\{n : S_n \leq t\}$.

We call $N(t)$ a *Poisson process with intensity* λ.

The intensity λ of a Poisson process can be taken to be any positive number. The choice very much depends on the application that we have in mind. For instance, when we think about earthquakes in California and take the unit of time as a year, then λ should not be so high, since the waiting times between successive eartquakes can be quite long. On the other hand, when we look at the moments that a radio-active material sends particles, with seconds as our time unit, then the intensity should be very high.

7.2   Basic Properties

In this section we prove a number of basic facts which should increase our understanding of the Poisson process, and which also justify the definition, given our objectives as mentioned at the beginning of the previous section. First of all, we should note that we have computed the distribution of S_n in Example 5.7.1. Indeed, the distribution of S_n, being the sum of n independent exponentially distributed random variables with parameter λ, is a *gamma distribution* and its density is given by

$$f_{S_n}(x) = \frac{\lambda^n}{(n-1)!} x^{n-1} e^{-\lambda x},$$

for $x \geq 0$, and $f_{S_n}(x) = 0$ for $x < 0$. We can use this to prove the following fact, which provides a link between Approach 1 and Approach 2 above.

Proposition 7.2.1. *We have*

$$P(N(t) = n) = \frac{(\lambda t)^n}{n!} e^{-\lambda t}.$$

Proof. The key to the proof is the observation that the event $N(t) \geq n$ is equivalent to $S_n \leq t$. Hence we can compute the distribution of $N(t)$ via the distribution of S_n, as follows:

$$
\begin{aligned}
P(N(t) = n) &= P(N(t) \geq n) - P(N(t) \geq n+1) \\
&= P(S_n \leq t) - P(S_{n+1} \leq t) \\
&= \int_0^t \lambda e^{-\lambda x} \frac{(\lambda x)^{n-1}}{(n-1)!} dx - \int_0^t \lambda e^{-\lambda x} \frac{(\lambda x)^n}{n!} dx.
\end{aligned}
$$

With partial integration, the first integral can be written as

$$\int_0^t \lambda e^{-\lambda x} \frac{(\lambda x)^{n-1}}{(n-1)!} dx = e^{-\lambda t} \frac{(\lambda t)^n}{n!} + \int_0^t \lambda e^{-\lambda x} \frac{(\lambda x)^n}{n!} dx,$$

which completes the proof. $\square$

The previous result tells us that the number of occurrences in the time interval $(0, t)$ has a Poisson distribution with parameter λt. The obvious question now is about the distribution of the number of occurrences in an arbitrary interval $(s, s + t)$. To study this, we need the following technical result:

Lemma 7.2.2. *For all $k \in \mathbb{N}$, $t > 0$, $u \in (0, t]$ and $v > 0$ we have*

$$P(t - u < S_k \leq t, t < S_{k+1} \leq t + v) = \frac{(\lambda t)^k - (\lambda(t - u))^k}{k!} e^{-\lambda t}(1 - e^{-\lambda v}).$$

Proof. Denote the joint density of S_k and X_{k+1} by f. By independence, we have

$$f(x, y) = f_{S_k}(x) f_{X_{k+1}}(y).$$

We can now write

$$P(t - u < S_k \leq t, t < S_{k+1} \leq t + v) = P(t - u < S_k \leq t, t < S_k + X_{k+1} \leq t + v)$$

$$= \int_{t-u}^{t} \int_{t-x}^{t+v-x} f(x, y) dy dx$$

$$= \int_{t-u}^{t} f_{S_k}(x) \left(\int_{t-x}^{t+v-x} f_{X_{k+1}}(y) dy \right) dx$$

$$= \int_{t-u}^{t} \frac{\lambda^k}{(k-1)!} x^{k-1} e^{-\lambda x} \times$$

$$\times (e^{-\lambda(t-x)} - e^{-\lambda(t+v-x)}) dx$$

$$= \frac{(\lambda t)^k - (\lambda(t - u))^k}{k!} e^{-\lambda t}(1 - e^{-\lambda v}). \qquad \square$$

Next, we fix some time t and suggest to study the *last* occurrence before t and the *first* occurrence after t. When $N(t) = k$, the last occurrence before t is S_k and the first occurrence after t is S_{k+1}. Hence, the last occurrence before time t can be written as $S_{N(t)}$. If there is no occurrence before time t, then $S_{N(t)} = S_0 = 0$ by convention. Similarly, $S_{N(t)+1}$ represents the first occurrence after time t.

♠ **Exercise 7.2.3.** Make sure you really understand this.

The time between t and $S_{N(t)+1}$ is not an ordinary waiting time; it is the *remainder* of the current waiting time at time t. You might expect that this remaining waiting time is in some sense 'smaller' than an 'ordinary' waiting time. However, the following surprising and fundamental result tells us that this is not the case. It tells us that the remaining waiting time at time t, and all subsequent waiting times are independent and have the same exponential distribution.

Theorem 7.2.4. *Let $V_t := S_{N(t)+1} - t$, and let for all $i = 2, 3, \ldots$, $Z_i := X_{N(t)+i}$. Then for all m, the random variables $V_t, Z_2, \ldots, Z_m$ are independent and identically distributed, with an exponential distribution with parameter λ.*

Proof. We express the distribution of V_t and the Z_i's in terms of the S_n's and X_n's as follows (using Lemma 7.2.2):

$$P(V_t \leq z_1, Z_2 \leq z_2, \ldots, Z_m \leq z_m)$$

$$= \sum_{k=0}^{\infty} P(N(t) = k, V_t \leq z_1, Z_2 \leq z_2, \ldots, Z_m \leq z_m)$$

$$= \sum_{k=0}^{\infty} P(N(t) = k, S_{k+1} - t \leq z_1, X_{k+2} \leq z_2, \ldots$$

$$\ldots, X_{k+m} \leq z_m)$$

$$= \sum_{k=0}^{\infty} P(S_k \leq t, t < S_{k+1} \leq t + z_1, X_{k+2} \leq z_2, \ldots$$

$$\ldots, X_{k+m} \leq z_m)$$

$$= \sum_{k=0}^{\infty} \frac{(\lambda t)^k}{k!} e^{-\lambda t} (1 - e^{-\lambda z_1}) \prod_{i=2}^{m} (1 - e^{-\lambda z_i})$$

$$= \prod_{i=1}^{m} (1 - e^{-\lambda z_i}) \sum_{k=0}^{\infty} P(N(t) = k)$$

$$= \prod_{i=1}^{m} (1 - e^{-\lambda z_i}),$$

proving the result.   □

The computations in the proof of Theorem 7.2.4 even show a little bit more:

Theorem 7.2.5. *For any m, the random variables $N(t), V_t, Z_2, \ldots, Z_m$ are independent.*

♠ **Exercise 7.2.6.** Show how Theorem 7.2.5 follows from the proof of Theorem 7.2.4.

We are now ready to consider $N(t + s) - N(t)$, which is the number of occurrences in an arbitrary interval $(t, t + s)$. The following theorem shows that this number also has a Poisson distribution, and that the number of occurrences in $(0, t)$ and $(t, t + s)$ are independent.

Theorem 7.2.7. *For all $t \geq 0$, $N(t + s) - N(t)$ has a Poisson distribution with parameter λs. Furthermore, $N(t)$ and $N(t + s) - N(t)$ are independent.*

Proof. Let $V_t, Z_2, Z_3, \ldots$ be defined as above, and write $Z_1 = V_t$. We observe that $N(t + s) - N(t)$ is equal to

$$N(t + s) - N(t) = \max\{i : Z_1 + Z_2 + \cdots + Z_i \leq s\}.$$

Hence the distribution of $N(t+s) - N(t)$ is fully determined by the joint distribution of V_t and the Z_i's. Since these are independent and exponentially distributed, the first claim follows immediately.

For the second claim, note that $N(t+s) - N(t)$ depends only on $V_t, Z_2, \ldots$, and by Lemma 7.2.4, these random variables are independent of $N(t)$ and the result follows. □

The independence result in the last theorem can be generalised considerably, as follows.

Theorem 7.2.8. *For any* $0 < t_1 < t_2 < \cdots < t_n$, *the random variables* $N(t_1)$, $N(t_2) - N(t_1)$, $N(t_3) - N(t_2), \ldots, N(t_n) - N(t_{n-1})$ *are independent.*

Proof. The proof proceeds by induction. For $n = 2$, the result is true according to Theorem 7.2.7. Now suppose that the result is true for $n = m$, and consider times $0 < t_1 < \cdots < t_{m+1}$. Let $V_t, Z_2, Z_3, \ldots$ be as defined before with t_1 instead of t, and let $T_i = V_t + Z_2 + \cdots + Z_i$, for all i. We use the notation $s_1 = t_2 - t_1, s_2 = t_3 - t_1, \ldots, s_m = t_{m+1} - t_1$. Then we can write

$$P(N(t_1) = k_1, N(t_2) - N(t_1) = k_2, \ldots, N(t_{m+1}) - N(t_m) = k_{m+1})$$
$$= P(N(t_1) = k_1, T_{k_2} \leq t_2 - t_1 < T_{k_2+1}, \ldots,$$
$$T_{k_2+\cdots+k_{m+1}} \leq t_{m+1} - t_m < T_{k_2+\cdots+k_{m+1}+1})$$
$$= P(N(t_1) = k_1)P(T_{k_2} \leq t_2 - t_1 < T_{k_2+1}, \ldots,$$
$$T_{k_2+\cdots+k_{m+1}} \leq t_{m+1} - t_m < T_{k_2+\cdots+k_{m+1}+1})$$
$$= P(N(t_1) = k_1)P(S_{k_2} \leq t_2 - t_1 < S_{k_2+1}, \ldots,$$
$$S_{k_2+\cdots+k_{m+1}} \leq t_{m+1} - t_m < S_{k_2+\cdots+k_{m+1}+1})$$
$$= P(N(t_1) = k_1)P(N(s_1) = k_2, N(s_2) - N(s_1) = k_3, \ldots,$$
$$N(s_m) - N(s_{m-1}) = k_{m+1})$$
$$= P(N(t_1) = k_1)P(N(s_1) = k_2) \cdots P(N(s_m) - N(s_{m-1}) = k_{m+1})$$
$$= P(N(t_1) = k_1)P(N(t_2) - N(t_1) = k_2) \cdots P(N(t_{m+1}) - N(t_m) = k_{m+1}),$$

proving the result. □

♠ **Exercise 7.2.9.** Justify all the equalities in the last proof. Where do we use the induction hypothesis?

At this point, it is perhaps good to look back at the points mentioned at the beginning of Section 7.1. We mentioned four characteristics of the type of processes that we want to model, namely stationarity, independence, lack of memory and no accumulation of occurrences. At this point we have addressed all these four issues. Indeed, stationarity follows from the fact that the distribution of $N(t+s) - N(t)$ does not depend on t, only on s, the length of the interval. Independence is addressed in Theorem 7.2.8, and lack of memory follows from Theorem 7.2.4. Finally, the fact that there is no accumulation of occurrences already follows from

Proposition 7.2.1. Indeed, since the number of occurrences in a bounded interval has a Poisson distribution, the probability to have infinitely many points in a given interval is equal to 0.

The Poisson process defined and studied so far seems to be a very reasonable model for the type of processes we have in mind. It is also a very interesting and subtle construction from a pure mathematical point of view, showing a nice interplay between discrete and continuous distributions. In the next sections, we shall explore some more of its properties.

7.3 The Waiting Time Paradox

The waiting times between successive occurrences have exponential distributions by construction. Nevertheless, there is something remarkable going on, as we can see when we take a close look at Theorem 7.2.4.

Fix a time t, and look at the waiting time between the last occurrence before t, and the first occurrence after t, that is, at

$$X_{N(t)+1} = S_{N(t)+1} - S_{N(t)}.$$

At first sight, this does not seem to be a very interesting question, since all X_i's have an exponential distribution. However, the point is that the index $N(t) + 1$ is *random*, and a little more thought reveals that random indices change the scene dramatically. To see that a random index can really make a big difference, consider the following example.

Define the random variable M as the first index i for which $X_i \geq 10$. It is clear that X_M does *not* have an exponential distribution since $P(X_M < 10) = 0$.

Hence, when we have random indices, the distribution might change. A look at Theorem 7.2.4 now does indeed tell us that $X_{N(t)+1}$ does *not* have an exponential distribution with parameter λ. To see this, note that

$$
\begin{aligned}
X_{N(t)+1} &= (S_{N(t)+1} - t) + (t - S_{N(t)}) \\
&= V_t + (t - S_{N(t)}).
\end{aligned}
$$

Since V_t itself has an exponential distribution with parameter λ, $X_{N(t)+1}$ should in some sense be *bigger* than an exponential distribution, since we add $t - S_{N(t)}$ as an extra factor. This fact is called the *waiting time paradox*.

It must be stressed that the waiting time paradox is not an artefact of a Poisson process, but a much more general phenomenon. To see this, *and* to get a better understanding of the paradox, we consider a very different situation in which the waiting times are not exponential.

Example 7.3.1. Consider random variables $X_1, X_2, \ldots$, independent and identically distributed with distribution

$$P(X = k) = \begin{cases} 1/2 & \text{for } k = 1, \\ 1/2 & \text{for } k = 100, \end{cases}$$

and interpret X_i as the waiting time between the $(i-1)$st and the ith occurrence. As before, we let $S_n = \sum_{i=1}^{n} X_i$ and $S_0 = 0$. Now let t be some large number and consider the length of the waiting time between the last occurrence before t and the first occurrence after t. What is the distribution of this quantity? Well, although individual waiting times are always equal to 100 with probability $1/2$ and equal to 1 with probability $1/2$, it is clear that the time periods with waiting time 100 cover a much higher fraction of the time axis than the time periods with waiting time 1. Therefore, it is much more likely that a given t is contained in a waiting time of length 100 than that it is contained in a waiting time of length 1. Hence, the expectation of $X_{N(t)+1}$ is certainly larger than $50\frac{1}{2}$. $\qquad \Box$

This example shows that the waiting time paradox is not restricted to Poisson processes, but is a general phenomenon, coming from the fact that the larger witing times cover a larger fraction of the time axis than the smaller ones.

In case of a Poisson process, we can do exact computations. We denote the time between t and the last occurrence before t by U_t, with the convention that if $N(t) = 0$, then $U_t = t$. The time between t and the next occurrence was denoted earlier by V_t. We know already from Theorem 7.2.4 that V_t has an exponential distribution with parameter λ.

Theorem 7.3.2. *The distribution of U_t is given by*

$$P(U_t \le u) = \begin{cases} 1 - e^{-\lambda u} & \text{for } 0 \le u < t, \\ 1 & \text{for } u \ge t. \end{cases}$$

Moreover, U_t and V_t are independent.

Proof. First of all, we have that

$$P(U_t = t, V_t \le v) = P(t < X_1 \le t + v) = e^{-\lambda t}(1 - e^{-\lambda v}). \qquad (7.4)$$

Furthermore, for $0 < u < t$ and $v > 0$ we find, using Lemma 7.2.2, that

$$
\begin{aligned}
P(U_t \le u, V_t \le v) &= P(S_{N(t)} \ge t - u, S_{N(t)+1} \le t + v) \\
&= \sum_{k=1}^{\infty} P(N(t) = k, S_k \ge t - u, S_{k+1} \le t + v) \\
&= \sum_{k=1}^{\infty} \frac{(\lambda t)^k - (\lambda(t-u))^k}{k!} e^{-\lambda t}(1 - e^{-\lambda v}) \\
&= (e^{\lambda t} - e^{\lambda(t-u)})(1 - e^{-\lambda v}).
\end{aligned}
$$

Sending $v \to \infty$ in (7.4) and the last expression yields the distribution of U_t as in the statement of the theorem. Independence also follows immediately from the given expression plus the fact that V_t has an exponential distribution with parameter λ. $\qquad \Box$

Note that U_t is a very natural example of a random variable which is neither discrete nor continuous; see Section 5.9 We can now compute the expectation of $X_{N(t)+1}$, representing the time between the last occurrence before t and the first occurrence after t.

Proposition 7.3.3. *It is the case that*

$$E(X_{N(t)+1}) = \frac{2}{\lambda} - \frac{e^{-\lambda t}}{\lambda},$$

and therefore, $E(X_{N(t)+1}) \to 2E(X_1) = \frac{2}{\lambda}$, as $t \to \infty$.

Proof. We have that $E(V_t) = \lambda^{-1}$. Furthermore,

$$
\begin{aligned}
E(U_t) &= \int_0^t x\lambda e^{-\lambda x}\,dx + te^{-\lambda t} \\
&= \lambda^{-1} - \lambda^{-1}e^{-\lambda t}.
\end{aligned}
$$

The result now follows immediately, since $X_{N(t)+1} = U_t + V_t$. □

All these results are quite intuitive. The distribution of U_t is an exponential distribution which is *truncated* at t because of the fact that at time t, the last occurrence before time t can not be more than t time units away. This truncation effect should disappear in the limit for $t \to \infty$, and indeed, the expectation of U_t does converge to the expectation of the exponential distribution with parameter λ.

7.4   The Strong Law of Large Numbers

In this section we give an interesting application of the strong law of large numbers. At time t we have seen $N(t)$ occurrences. The average waiting time between two occurrences is equal to λ^{-1}, the expectation of an individual X_i. Hence in a time interval of length t, we should expect around λt occurrences, that is, we expect $N(t)$ te be close to λt.

Theorem 7.4.1 (Strong law of large numbers). *Let $\epsilon > 0$. Then*

$$P\left(\lim_{t\to\infty} \frac{N(t)}{t} = \lambda\right) = 1.$$

Proof. Observe that

$$S_{N(t)} \le t \le S_{N(t)+1}, \tag{7.5}$$

and hence division by $N(t)$ gives

$$\frac{1}{N(t)}\sum_{i=1}^{N(t)} X_i \le \frac{t}{N(t)} \le \frac{1}{N(t)}\sum_{i=1}^{N(t)+1} X_i. \tag{7.6}$$

According to Theorem 6.3.2, we have that the left and right hand side of (7.6) converge to λ^{-1}, with probability 1. It follows immediately that also $t/N(t)$ converges to λ^{-1}, with probability 1. $\qquad\square$

This law of large numbers can be used for statistical applications. If you have some process that you want to decsribe with a Poisson process, then you may want to get an idea about a suitable intensity λ. The last result tells us that $N(t)/t$ should with high probability be close to λ. When we *observe* $N(t)$, we can *estimate* λ using this fact.

♠ **Exercise 7.4.2.** Do you think that is easier or more difficult to estimate λ when λ gets larger? Motivate your answer.

7.5 Exercises

Exercise 7.5.1. An employee in a call center works form 8 a.m. until 5 p.m., with breaks between 10.30-10.45, 12.30-13.30 and 14.45-15.00. Assume that calls come in according to a Poisson process with expected number of calls per hour equal to 6.
(a) What is the probability that there are at most 10 calls during the breaks?
(b) What is the probability that the first call of the day is after 8.10 a.m.?
(c) What is the probability that the employee can do something else for 45 minutes without being disturbed by a call?

Exercise 7.5.2. Let $0 < s < t$. Compute the conditional distribution of $N(s)$ given that $N(t) = n$. Do you recognize this distribution?

Exercise 7.5.3. Compute the distribution of $S_{N(t)}$.

Exercise 7.5.4. Consider a Poisson process with parameter λ. What is the conditional probability that $N(1) = n$ given that $N(3) = n$? Do you understand why this probability does not depend on λ?

Exercise 7.5.5. Give an alternative proof of Lemma 7.2.2, using Theorem 7.2.8.

Exercise 7.5.6. One can compute the exact distribution of $X_{N(t)+1}$, the time interval between the last occurrence before time t and the first occurrence after time t. Show that the density f of $X_{N(t)+1}$ is given by

$$f(x) = \begin{cases} \lambda^2 x e^{-\lambda x} & \text{for } x < t, \\ \lambda(1 + \lambda t)e^{-\lambda x} & \text{for } x > t, \end{cases}$$

and compute the expectation of $X_{N(t)+1}$ using this density.

Exercise 7.5.7 (Thinning a Poisson process). Let $N(t)$ be a Poisson process with intensity λ. For each occurrence, we flip a coin: if heads comes up we label the occurrence *green*, if tails comes up we label it *red*. The coin flips are independent and p is the probability to see heads.

(a) Show that the green occurrence form a Poisson process with intensity λp.

(b) Connect this with Example 2.2.5.

(c) We claim that the red occurrences on the one hand, and the green occurrences on the other hand form *independent* Poisson processes. Can you formulate this formally, and prove it , using Example 2.2.5 once more?

Exercise 7.5.8 (The inhomogeneous Poisson process). In the theory of this chapter, the intensity λ was a constant. However, there might be situations in which it is more reasonable to allow for a varying intensity of the process. Instead of a fixed intensity λ, we want to have a model in which $\lambda = \lambda(t)$, a function of the time t.

(a) Can you think of a number of examples where this might be a reasonable thing to do?

(b) Define such an *inhomogeneous* Poisson process, and show that (under certain conditions) the distribution of the number of occurrences in the time interval (a, b) is given by a Poisson distribution with parameter $\int_a^b \lambda(t)dt$.

Exercise 7.5.9 (The Poisson process in higher dimensions). We can extend the idea of random occurrences of events to higher dimensions. Of course, we should not talk about occurrences in *time* anymore then, but instead talk about objects that we distribute over some two-dimensional set in some completely random way. Think of how you would define such a two-dimensional process. In particular, note that not all approaches to the Poisson process in this chapter can be generalised to higher dimensions. For instance, the waiting time does not have an interpretation in two dimensions. But the number of points in a given region can possibly again be described by a Poisson distribution. Conclude that in generalising the process to higehr dimensions, not all approaches are equally suitable.

Chapter 8

Limit Theorems

In this chapter, we will be concerned with some more general limit theorems. In particular, we shall generalise the central limit Theorem 4.2.1. The method of proof will also lead to a new formulation of the law of large numbers. The methods behind these results are not so easy. They rely on concepts from complex analysis. To make sure that that you know what we are talking about, in Section 8.2 there will be a short introduction to complex analysis which contains all the background necessary for the development in this chapter. In this chapter, random variables can be discrete or continuous. We start by formalising a mode of convergence that we have, in fact, already seen.

8.1 Weak Convergence

When we have random variables X and X_n, there are several ways of expressing the idea that X_n should be close to X. For instance, when the random variables are defined on the same sample space, we can look at $P(|X_n - X| > \epsilon)$. This was the kind of 'closeness' that was used in the weak law of large numbers Theorem 4.1.4, for instance. We want to emphasize that this type of closeness only makes sense when all random variables are defined on the same sample space.

In this chapter, we will discuss another type of closeness, which does not require the random variables to be defined on the same sample space. The type of convergence that we have in mind is called *weak convergence* and is defined via the distribution function of the random variables. We think it is best to first give the formal definition and comment on the definition afterwards.

Definition 8.1.1. Let $X, X_1, X_2, \ldots$ be random variables with distribution functions $F, F_1, F_2, \ldots$ respectively. We say that the sequence $X_1, X_2, \ldots$ *converges weakly to X* if

$$F_n(x) \to F(x),$$

for all x in which F is continuous, and we write this as

$$X_n \Rightarrow X \text{ or } F_n \Rightarrow F.$$

Weak convergence is also called *convergence in distribution* in the literature. At first sight, this definition may appear somewhat unnatural, especially the fact that we do not require convergence in points of discontinuity of F. At this point, we cannot fully explain this, but the following example does indicate the reason for the fact that convergence is not required at points of discontinuity of F.

Example 8.1.2. Let F_n be defined by

$$F_n(x) = \begin{cases} 0 & \text{if } x < \frac{1}{n}, \\ 1 & \text{if } x \geq \frac{1}{n}. \end{cases}$$

F_n is the distribution function of a random variable X_n which satisfies $P(X_n = 1/n) = 1$. Clearly, when $n \to \infty$, the distribution of X_n gets 'closer' to the distribution of a random variable X with $P(X = 0) = 1$. However, if we denote the distribution function of X by F, then we see that $F_n(0)$ does not converge to $F(0)$, as $n \to \infty$. Indeed, $F_n(0) = 0$ for all n, but $F(0) = 1$. □

♠ **Exercise 8.1.3.** Let F'_n be defined by

$$F'_n(x) = \begin{cases} 0 & \text{if } x < -\frac{1}{n}, \\ 1 & \text{if } x \geq -\frac{1}{n}. \end{cases}$$

Show that in this case, $F'_n(x) \to F(x)$ for *all* x.

The distribution functions F'_n correspond to random variables Y_n which satisfy $P(Y_n = -\frac{1}{n}) = 1$. Hence we see that if we had required convergence for all x, the random variables X_n would not converge weakly to X, but the Y_n would. This would be a somewhat strange state of affairs, which is avoided by the way we defined weak convergence. We do realise though, that this does not fully account for the definition, and at this point We just would like to ask for some patience.

Example 8.1.4. The Central Limit Theorem 4.2.1 is an example of weak convergence. □

The following example shows that it is possible that discrete random variables converge weakly to a continuous random variable.

Example 8.1.5. Let X_n be a random variable taking the values $\{1/n, 2/n, \ldots, 1\}$, each with probability $1/n$, and let X be a uniformly distributed random variable on $(0, 1)$. Denote the corresponding distribution functions by F_n and F, respectively. We claim that $X_n \Rightarrow X$. To see this, take $0 \leq y \leq 1$, and observe that

$$P(X_n \leq y) = \frac{\lfloor ny \rfloor}{n},$$

where $\lfloor x \rfloor$ denotes the largest integer which is smaller than or equal to x. Clearly, the right hand side converges to y, as $n \to \infty$. For $y < 0$ and $y > 1$, we have $F_n(y) = 0$ and $F_n(y) = 1$, respectively. It is now clear that $F_n \Rightarrow F$. □

In the following exercise it is shown that it is possible for continuous random variables $X, X_1, X_2, \ldots$, that $X_n \Rightarrow X$ in such a way that the densities do not converge.

♠ **Exercise 8.1.6.** Let X_n have distribution function

$$F_n(x) = x - \frac{\sin(2n\pi x)}{2n\pi},$$

for $0 \le x \le 1$. First show that F_n is indeed a distribution function and after that, show that X_n converges weakly to the uniform distribution on $(0, 1)$, but that the density of X_n does not converge to the density of the uniform distribution.

8.2 Characteristic Functions

Characteristic functions are one of the main tools in studying weak convergence. In this section, we introduce these characteristic functions and derive some of their basic properties. Unfortunately, characteristic functions require a certain amount of complex analysis, not much, but perhaps just enough to scare you off. Don't worry too much about this, we will define all the necessary machinery in the course of the developments, and some results from complex analysis we will just take for granted.

First of all, we shall need the exponential function

$$e^z = \sum_{k=0}^{\infty} \frac{z^k}{k!},$$

for all $z \in \mathbb{C}$. This function has the nice property of being its own derivative. The standard limit

$$\lim_{n\to\infty} \left(1 + \frac{z}{n}\right)^n = e^z, \tag{8.1}$$

which we recognise from real analysis, remains valid in the complex setting. We also note the important identity

$$e^{it} = \cos t + i \sin t.$$

For a continuous function $f : (a, b) \to \mathbb{C}$, we define

$$\int_a^b f(t)dt = \int_a^b \mathrm{Re} f(t)dt + i \int_a^b \mathrm{Im} f(t)dt,$$

where $\mathrm{Re} f$ and $\mathrm{Im} f$ are the real and imaginary part of f, respectively. In particular we find that

$$\int_a^b e^{it}dt = \int_a^b \cos t\, dt + i \int_a^b \sin t\, dt.$$

In view of this, it makes sense to define, for a random variable X,

$$E(e^{itX}) := E(\cos tX) + iE(\sin tX).$$

This expectation is called the *characteristic function* of X. Characteristic functions are strongly related to the generating functions in Section 2.6.

Definition 8.2.1. Let X be a random variable. The *characteristic function* of X, denoted by $\phi_X(t) : \mathbb{R} \to \mathbb{C}$, is defined as

$$\phi_X(t) = E(\cos tX) + iE(\sin tX),$$

which we often denote by $E(e^{itX})$.

So, if X is a continuous random variable, then

$$\phi_X(t) = \int_{-\infty}^{\infty} f_X(x)e^{itx}dx,$$

and if X is discrete, we obtain

$$\phi_X(t) = \sum_x e^{itx} P(X = x).$$

Working with characteristic functions requires some knowledge about complex integrals. Some basic properties of these integrals are given in the next theorem, where f and g are complex-valued functions on $\mathbb{R}$.

Theorem 8.2.2.   (a) $\int_a^b (f + g)(t)dt = \int_a^b f(t)dt + \int_a^b g(t)dt$.

(b) $\int_a^b (\lambda f)(t)dt = \lambda \int_a^b f(t)dt$, *for all $\lambda \in \mathbb{C}$.*

(c) $\left| \int_a^b f(t)dt \right| \leq \int_a^b |f(t)|\, dt$.

Proof. (a) follows from the corresponding properties of real integrals after splitting the integral in a real and imaginary part. The details are left as an exercise. For (b), it is convenient to write $\lambda = \lambda_1 + i\lambda_2$, with λ_1 and λ_2 in $\mathbb{R}$, and also write $f = f_1 + if_2$, with f_1 and f_2 the real and imaginary part of f respectively. We then write

$$
\begin{aligned}
\int_a^b (\lambda f)(t)dt &= \lambda_1 \int_a^b f_1(t)dt - \lambda_2 \int_a^b f_2(t)dt \\
&\quad + i\lambda_2 \int_a^b f_1(t)dt + i\lambda_1 \int_a^b f_2(t)dt \\
&= (\lambda_1 + i\lambda_2) \int_a^b (f_1 + if_2)(t)dt = \lambda \int_a^b f(t)dt.
\end{aligned}
$$

The proof of (c) is more subtle. We first write $\int_a^b f(t)dt = re^{i\theta}$, where r is the modulus, and θ the argument, as usual. Then we have

$$r = e^{-i\theta} \int_a^b f(t)dt = \int_a^b e^{-i\theta} f(t)dt.$$

Since $r \in \mathbb{R}$ it now follows that the last integral is real, and is therefore equal to its real part. Hence we obtain that

$$
\begin{aligned}
\left| \int_a^b f(t)dt \right| &= \left| e^{-i\theta} \int_a^b f(t)dt \right| = r \\
&= \int_a^b e^{-i\theta} f(t)dt = \int_a^b \mathrm{Re}(e^{-i\theta} f(t))dt \\
&\leq \int_a^b \left| e^{-i\theta} f(t) \right| dt \\
&= \int_a^b |f(t)|\, dt.
\end{aligned}
$$

$\square$

Before giving some examples of characteristic functions, we would like to mention (and prove) two important properties.

Theorem 8.2.3. (a) *If X and Y are independent, then*

$$\phi_{X+Y}(t) = \phi_X(t)\phi_Y(t).$$

(b) *If $a, b \in \mathbb{R}$, and $Y = aX + b$, then*

$$\phi_Y(t) = e^{itb}\phi_X(at).$$

Proof. For (a), we can use Theorem 5.6.13. Hence, we should again split everything in real and imaginary parts as follows.

$$
\begin{aligned}
\phi_{X+Y}(t) &= E(e^{it(X+Y)}) = E(e^{itX}e^{itY}) \\
&= E((\cos tX + i\sin tX)(\cos tY + i\sin tY)) \\
&= E(\cos tX \cos tY - \sin tX \sin tY) + \\
&\quad +iE(\sin tX \cos tY + \cos tX \sin tY) \\
&= E(\cos tX)E(\cos tY) - E(\sin tX)E(\sin tY) + \\
&\quad +i(E(\sin tX)E(\cos tY) + E(\cos tX)E(\sin tY)) \\
&= (E(\cos tX) + iE(\sin tX))(E(\cos tY) + iE(\sin tY)) \\
&= \phi_X(t)\phi_Y(t).
\end{aligned}
$$

For (b), we write

$$
\begin{aligned}
\phi_Y(t) &= E(e^{it(aX+b)}) = E(e^{itb}e^{i(at)X}) \\
&= e^{itb}E(e^{i(at)X}) = e^{itb}\phi_X(at).
\end{aligned}
$$

$\square$

♠ **Exercise 8.2.4.** Extend the first part of this theorem to more than two random variables.

Example 8.2.5. Let X be a discrete random variable with $P(X = 1) = p$ and $P(X = 0) = 1 - p$. Then $\phi_X(t) = (1 - p) + pe^{it}$.   □

Example 8.2.6. If X has a binomial distribution with parameters n and p, then X can be written as the sum of n independent random variables as in the previous example. Hence, according to Theorem 8.2.3, we find that

$$\phi_X(t) = (1 - p + pe^{it})^n.$$ □

♠ **Exercise 8.2.7.** Let X have an exponential distribution with parameter 1. Then its characteristic function is $\phi_X(t) = \int_0^\infty e^{itx} e^{-x} dx$. In this case, we can compute this by repeated partial integration. For this, we first consider the real part $\int_0^\infty \cos tx e^{-x} dx$. Compute this integral by doing partial integration twice. Do the same for the imaginary part, put things together and show that

$$\phi_X(t) = \frac{1}{1 - it}.$$

♠ **Exercise 8.2.8.** Show that when X has a Poisson distribution with parameter λ, we have

$$\phi_X(t) = e^{-\lambda(1 - e^{it})}.$$

These examples suggest that we can typically compute characteristic functions without too much trouble. Unfortunately, this is not the case. For instance, at this point we cannot compute the characteristic function of the normal distribution or the Cauchy distributions. We will come back to this soon.

We think it is about time to explain why we should be interested in characteristic functions at all. Two major results form the basis for this interest, the *inversion theorem* and the *continuity theorem*. The inversion theorem explains the name *characteristic* function. Indeed, it tells us that the distribution of a random variable is completely determined by its characteristic function.

Theorem 8.2.9 (Inversion theorem). (a) *Two random variables X and Y have the same characteristic function if and only if they have the same distribution function.*

(b) *If X is continuous with density f and characteristic function ϕ, then*

$$f(x) = \frac{1}{2\pi} \int_{-\infty}^\infty e^{-itx} \phi(t) dt,$$

at every point x at which f is differentiable.

(c) *If X takes values in $\mathbb{N}$ and has characteristic function ϕ, then*

$$P(X = k) = \frac{1}{2\pi} \int_{-\pi}^\pi e^{-itk} \phi(t) dt.$$

Proof. We will not give a proof of (a) here. We have two reasons for that. Firstly, the general proof is somewhat technical and perhaps not so suitable for a book at this level. Secondly (and more importantly), in this book we mainly use (b) and (c), which are refinements in two special cases of (a). The statement in (b) is nothing but the classical Fourier inversion theorem, and the proof can be found in any introduction to Fourier transforms. This is also true for the proof of (c), but since it is so simple, we prefer to give it here. The proof of (c) runs as follows. It is an easy exercise to show that

$$\int_{-\pi}^{\pi} e^{itk} dt = 0,$$

when $k \neq 0$, and of course equal to 2π when $k = 0$. Hence we can write

$$P(X = k) = \frac{1}{2\pi} \sum_{j=-\infty}^{\infty} P(X = j) \int_{-\pi}^{\pi} e^{itj} e^{-itk} dt$$

$$= \frac{1}{2\pi} \int_{-\pi}^{\pi} e^{-itk} \phi(t) dt,$$

proving the result. □

♠ **Exercise 8.2.10.** The careful reader will have noticed that in the last proof, we interchanged sum and integral. Do you see why this is allowed?

Finally, the following *continuity theorem* articulates the relation between weak convergence and characteristic functions.

Theorem 8.2.11 (Continuity theorem). *Suppose that $X, X_1, X_2, \ldots$ are random variables with characteristic functions $\phi, \phi_1, \phi_2, \ldots$. Then*

$$X_n \Rightarrow X$$

if and only if

$$\phi_n(t) \to \phi(t),$$

for all t, when $n \to \infty$.

Perhaps it now becomes clear how we are going to prove limit theorems. What we need to do is to compute characteristic functions, and identify limits. This is easier said than done. In the next section, we will demonstrate an expansion of the characteristic functions which turns out to be useful for both purposes: computing characteristing functions, and computing limits.

8.3 Expansion of the Characteristic Function

This section is rather technical in nature, but we want to emphasise that the computations are not very difficult. If you get tired of the computational details, then it will be enough to study the results without looking at the proofs. In fact, the proofs do not give much information at all, but we prefer to include them.

Everything is based on the following estimate.

Lemma 8.3.1. *For all $x \in \mathbb{R}$ and $n \geq 0$ we have*

$$\left| e^{ix} - \sum_{k=0}^{n} \frac{(ix)^k}{k!} \right| \leq \min \left\{ \frac{|x|^{n+1}}{(n+1)!}, \frac{2|x|^n}{n!} \right\}.$$

Proof. The proof is based on partial integration in the complex setting. We haven't told you before that you are allowed to do that, but in fact the usual partial integration formula (under certain continuity assumptions) holds.

The first step of the proof is to show by partial integration that

$$\int_0^x (x-s)^n e^{is} ds = \frac{x^{n+1}}{n+1} + \frac{i}{n+1} \int_0^x (x-s)^{n+1} e^{is} ds. \tag{8.2}$$

From this formula it is not hard to show by induction that

$$e^{ix} = \sum_{k=0}^{n} \frac{(ix)^k}{k!} + \frac{i^{n+1}}{n!} \int_0^x (x-s)^n e^{is} ds. \tag{8.3}$$

Hence,

$$\left| e^{ix} - \sum_{k=0}^{n} \frac{(ix)^k}{k!} \right| \quad = \quad \left| \frac{i^{n+1}}{n!} \int_0^x (x-s)^n e^{is} ds \right|$$

$$\leq \quad \frac{1}{n!} \int_0^x (x-s)^n ds$$

$$= \quad \frac{|x|^{n+1}}{(n+1)!},$$

accounting for the first term at the right hand side of the lemma.

Substituting $n-1$ for n in (8.2) gives

$$\int_0^x (x-s)^n e^{is} ds = \frac{n}{i} \int_0^x (x-s)^{n-1} e^{is} ds - \frac{x^n}{i},$$

and substituting this into (8.3) tells us that

$$e^{ix} \quad = \quad \sum_{k=0}^{n} \frac{(ix)^k}{k!} + \frac{i^{n+1}}{n!} \left(\frac{n}{i} \int_0^x (x-s)^{n-1} e^{is} ds - \frac{x^n}{i} \right)$$

$$= \quad \sum_{k=0}^{n} \frac{(ix)^k}{k!} + \frac{i^n}{(n-1)!} \int_0^x (x-s)^{n-1} (e^{is} - 1) ds.$$

We can now estimate the integral at the right hand side (using that $|e^{is} - 1| \leq 2$) in the same way as above, giving the second term at the right hand side in the lemma. $\square$

Having done this unattractive and somewhat tedious computation, we can now enjoy the corollaries:

Theorem 8.3.2. *Let X be a random variable for which $E(|X|^n) < \infty$. Then,*

$$\phi_X(t) = \sum_{k=0}^{n} \frac{(it)^k}{k!} E(X^k) + \beta(t),$$

where $\beta(t)/t^n \to 0$ as $t \to 0$.

Proof. We give the proof for continuous random variables.

$$
\begin{aligned}
\left| \phi_X(t) - \sum_{k=0}^{n} \frac{(it)^k}{k!} E(X^k) \right|
&= \left| \int_{-\infty}^{\infty} \left(e^{itx} - \sum_{k=0}^{n} \frac{(it)^k}{k!} x^k \right) f_X(x) dx \right| \\
&\le \int_{-\infty}^{\infty} \left| e^{itx} - \sum_{k=0}^{n} \frac{(it)^k}{k!} x^k \right| f_X(x) dx \\
&\le \int_{-\infty}^{\infty} \min \left\{ \frac{|tx|^{n+1}}{(n+1)!}, \frac{2|tx|^n}{n!} \right\} f_X(x) dx \\
&= \frac{t^n}{n!} \int_{-\infty}^{\infty} \min \left\{ \frac{t|x|^{n+1}}{(n+1)}, 2|x|^n \right\} f_X(x) dx.
\end{aligned}
$$

The minimum in the last formula is clearly bounded above by $f_X(x)2|x|^n$ and by assumption, the integral of this last function is finite. Since the first term inside the min goes to 0 as $t \to 0$, the integrand tends to zero, and by the dominated convergence Theorem 3.2.4, the whole integral tends to 0. This proves the result. $\square$

The next result can be used to compute the characteristic function of a normal random variable.

Theorem 8.3.3. *Let X have characteristic function ϕ, and let t be such that*

$$\lim_{n \to \infty} \frac{t^n E(|X|^n)}{n!} = 0. \tag{8.4}$$

Then $\phi(t)$ has the expansion

$$\phi(t) = \sum_{k=0}^{\infty} \frac{(it)^k}{k!} E(X^k).$$

♠ **Exercise 8.3.4.** Prove this result, using the proof of the previous theorem.

Here follows a nice application of the last theorem.

Example 8.3.5. In this example, we compute the characteristic function of the normal distribution. In Section 8.5 we shall see how important this computation is.

Let X have a standard normal distribution. First, we want to compute $E(X^k)$. It is easy to see that this is zero for odd k, and we therefore concentrate on even k. Integration by parts shows that

$$\frac{1}{\sqrt{2\pi}} \int_{-\infty}^{\infty} x^k e^{-x^2/2} dx = \frac{k-1}{\sqrt{2\pi}} \int_{-\infty}^{\infty} x^{k-2} e^{-x^2/2} dx.$$

(To do this partial integration, observe that the integrand on the left can be written as $x^{k-1} \cdot x e^{-x^2/2}$.) Now $E(X^0) = 1$, and the formula tells us that $E(X^{2k}) = (2k-1)E(X^{2k-2})$. Hence we find that

$$E(X^{2k}) = 1 \cdot 3 \cdot 5 \cdots (2k-1).$$

From this it follows that (8.4) is satisfied for all t, and Theorem 8.3.3 applies. It follows that

$$
\begin{aligned}
\phi_X(t) &= \sum_{k=0}^{\infty} \frac{(it)^{2k}}{(2k)!} 1 \cdot 3 \cdot 5 \cdots (2k-1) \\
&= \sum_{k=0}^{\infty} (-t^2)^k \frac{1 \cdot 3 \cdots (2k-1)}{1 \cdot 2 \cdot 3 \cdots 2k} \\
&= \sum_{k=0}^{\infty} (-t^2)^k \frac{1}{2^k k!} = e^{-t^2/2}.
\end{aligned}
$$

$\square$

♠ **Exercise 8.3.6.** Compute, using Theorem 8.2.3 and Example 5.3.6, the characteristic function of a random variable with normal distribution with parameters μ and σ^2.

8.4 The Law of Large Numbers

Now we have set up things in such a way that we can state and quickly prove our next law of large numbers. This law will be in terms of weak convergence. The main reason to include this law is to illustrate a certain technique. After stating and proving the result, we will compare this law of large numbers to Theorem 4.1.4.

Theorem 8.4.1 (Law of large numbers). *Let* $X_1, X_2, \ldots$ *be independent random variables with the same distribution, and suppose that they have finite expectation* μ. *Let* $S_n = X_1 + \cdots + X_n$. *Then*

$$\frac{S_n}{n} \Rightarrow \mu,$$

that is, $P(S_n/n \leq x) \to 0$ *if* $x < \mu$ *and* $P(S_n/n \leq x) \to 1$ *if* $x > \mu$.

Before proving this result, let me state a small technical lemma that turns out to be useful in the proof.

Lemma 8.4.2. *Let $z_1, \ldots, z_m$ and $z'_1, \ldots, z'_m$ be complex numbers of modulus at most r. Then*

$$|z_1 z_2 \cdots z_m - z'_1 z'_2 \cdots z'_m| \le r^{m-1} \sum_{k=1}^{m} |z_k - z'_k|.$$

Proof. This can be proved by induction. For $m = 1$ the result is clearly true. Suppose that the result is true for $m - 1$. Note the following identity:

$$z_1 z_2 \cdots z_m - z'_1 z'_2 \cdots z'_m = (z_1 - z'_1)(z_2 z_3 \cdots z_m) + z'_1(z_2 z_3 \cdots z_m - z'_2 z'_3 \cdots z'_m).$$

Hence, we have that

$$
\begin{aligned}
|z_1 \cdots z_m - z'_1 \cdots z'_m| &\le |(z_1 - z'_1)(z_2 \cdots z_m)| + |z'_1(z_2 \cdots z_m - z'_2 \cdots z'_m)| \\
&\le r^{m-1}|z_1 - z'_1| + r|z_2 \cdots z_m - z'_2 \cdots z'_m| \\
&\le r^{m-1} \sum_{k=1}^{m} |z_k - z'_k|,
\end{aligned}
$$

according to the induction hypothesis. □

Proof of Theorem 8.4.1. The method of proof is perhaps clear by now. We need to show, by the Continuity Theorem 8.2.11, that the characteristic function of S_n/n converges to the characteristic function of the constant μ. The characteristic function of S_n/n can be computed with the help of Theorem 8.3.2 as follows.

Let $\phi(t)$ be the characteristic function common to the X_i's. The characteristic function of $X_1 + \cdots + X_n$ is then equal to $\phi(t)^n$, according to Theorem 8.2.3. According to the same theorem, the characteristic function of S_n/n is then equal to $\phi(t/n)^n$. According to Theorem 8.3.2, as $n \to \infty$ (which implies that $t/n \to 0$), we have

$$\phi\left(\frac{t}{n}\right)^n = \left(1 + \frac{i\mu t}{n} + \beta\left(\frac{t}{n}\right)\right)^n,$$

where $\beta(t/n)/(t/n) \to 0$, as $n \to \infty$. It then follows that

$$n\left|\phi\left(\frac{t}{n}\right) - \left(1 + \frac{i\mu t}{n}\right)\right| = \beta\left(\frac{t}{n}\right) n \to 0, \tag{8.5}$$

as $n \to \infty$ by the fact that $\beta(t/n)/(t/n) \to 0$. Now we apply Lemma 8.4.2 with $m = n$, $z_k = 1 + (i\mu t)/n$, $z'_k = \phi(t/n)$ and $r = 1 + \mu t/n$. It then follows that

$$\left|\phi\left(\frac{t}{n}\right)^n - \left(1 + \frac{i\mu t}{n}\right)^n\right| \le \left(1 + \frac{\mu t}{n}\right)^{n-1} n \left|\phi\left(\frac{t}{n}\right) - \left(1 + \frac{i\mu t}{n}\right)\right|,$$

which goes to 0, using (8.1) and (8.5). Since $(1 + (i\mu t)/n)^n \to e^{i\mu t}$ (according to (8.1)), it follows that also $\phi(t/n)^n \to e^{i\mu t}$, which is the characteristic function of the constant μ. □

♠ **Exercise 8.4.3.** It is not so easy to compute the characteristic function of the Cauchy distribution. One needs so called *contour integrals* for this. But let me tell you that this characteristic function is

$$\phi(t) = e^{-|t|}.$$

Let $X_1, \ldots, X_n$ be independent, all having a Cauchy distribution. Show that $(X_1 + \cdots + X_n)/n$ again has a Cauchy distribution. Why does this not contradict Theorem 8.4.1?

It is now interesting to compare Theorem 4.1.4 to Theorem 8.4.1. There is a sense in which the last theorem is in fact a consequence of the first. This follows from the following general result, which compares two different modes of convergence.

Theorem 8.4.4. *Let $X, X_1, X_2, \ldots$ be defined on the same sample space. If*

$$P(|X_n - X| \geq \epsilon) \to 0,$$

for all $\epsilon > 0$, then

$$X_n \Rightarrow X.$$

Proof. Denote the distribution functions of X and X_n by F and F_n respectively. Then we write, for any $\epsilon > 0$,

$$
\begin{aligned}
F_n(x) &= P(X_n \leq x) \\
&= P(X_n \leq x, X \leq x + \epsilon) + P(X_n \leq x, X > x + \epsilon) \\
&\leq P(X \leq x + \epsilon) + P(|X - X_n| > \epsilon) \\
&= F(x + \epsilon) + P(|X - X_n| > \epsilon).
\end{aligned}
$$

Similarly we have

$$
\begin{aligned}
F(x - \epsilon) &= P(X \leq x - \epsilon) \\
&= P(X \leq x - \epsilon, X_n \leq x) + P(X \leq x - \epsilon, X_n > x) \\
&\leq P(X_n \leq x) + P(|X - X_n| > \epsilon) \\
&= F_n(x) + P(|X - X_n| > \epsilon).
\end{aligned}
$$

When we combine these two estimates, we obtain

$$F(x - \epsilon) - P(|X - X_n| > \epsilon) \leq F_n(x) \leq F(x + \epsilon) + P(|X - X_n| > \epsilon).$$

Now let $n \to \infty$. We then obtain

$$F(x - \epsilon) \leq \liminf_{n \to \infty} F_n(x) \leq \limsup_{n \to \infty} F_n(x) \leq F(x + \epsilon).$$

Now assume that x is a continuity point of F. In that case, sending $\epsilon \to 0$ leads to $F_n(x) \to F(x)$. Hence we have shown pointwise convergence of F_n to F in continuity points of F, and the proof is complete. □

8.5 The Central Limit Theorem

I was once told that the central limit theorem is the most quoted result in *mathematics*, not only in probability. Whether this is true or not I do not pretend to know, but it is certainly true that the central limit theorem does play a crucial role in probability.

In this introductory text, we will only state and prove the most basic central limit theorem. This result expresses the idea that when you take the sum of many independent random variables with the same distribution, this sum approximately has a normal distribution, regardless of the distribution of the summands! This is quite a startling fact: one can take Poisson random variables, add them up, and the result will be roughly normally distributed. If one starts with exponential random variables, the conclusion will be the same. Here is the formal result.

Theorem 8.5.1 (Central limit theorem). *Let $X_1, X_2, \ldots$ be independent random variables with the same distribution, and suppose that their common expectation μ and variance σ^2 are both finite. Let $S_n = X_1 + \cdots + X_n$. Then*

$$\frac{S_n - n\mu}{\sigma\sqrt{n}} \Rightarrow N,$$

where N denotes a random variable with a standard normal distribution.

Proof. The idea of the proof is the same as the proof of Theorem 8.4.1. Writing

$$Y_i = \frac{X_i - \mu}{\sigma},$$

we see that $E(Y_i) = 0$ and $E(Y_i^2) = 1$. We write ϕ for the characteristic function of the Y_i's. It follows from Theorem 8.3.2 that

$$\phi(t) = 1 - \frac{1}{2}t^2 + \beta(t),$$

where $\beta(t)/t^2 \to 0$ as $t \to 0$. Since

$$\frac{S_n - n\mu}{\sigma\sqrt{n}} = \frac{1}{\sqrt{n}}\sum_{i=1}^{n} Y_i, \tag{8.6}$$

we see from Theorem 8.2.3 that the characteristic function of (8.6) is given by

$$\phi\left(\frac{t}{\sqrt{n}}\right)^n = \left(1 - \frac{t^2}{2n} + \beta\left(\frac{t}{\sqrt{n}}\right)\right)^n$$

The proof is now finished in exactly the way as the proof of Theorem 8.4.1. □

♠ **Exercise 8.5.2.** Provide the missing details in this proof.

The following example is a typical statistical application of the central limit theorem.

Example 8.5.3. It is conjectured that one can see from a person's handwriting, whether this person is a man or a woman. Suppose that someone inspects 1500 examples of handwriting and that he or she assigns the right sex to the handwriting 950 times. Is the conjecture plausible and why?

To say something about this, imagine that there would be no difference in general between the handwriting of men and women. In that case, it is only reasonable that out of 1500 trials, the number of correct guesses would be roughly 750. So in that sense, 950 correct guesses seems to be evidence for the idea that there is, in fact, a difference. A standard way to quantify this, is to ask the question how likely it would be, under the condition that there is *no* difference, to make *at least* 950 correct guesses. (Clearly the probability to make *exactly* 950 correct guesses would be very small.) If we assume that there is no difference, the number of correct guesses can perhaps be written as $X = \sum_{i=1}^{1500} X_i$, where $P(X_i = 1) = P(X_i = 0) = \frac{1}{2}$. Hence this number of correct guesses is a sum of 1500 independent random variables with the same distribution, and the central limit theorem applies. The expectation of the X_i's is $\frac{1}{2}$ and the variance is $\frac{1}{4}$. Hence

$$\frac{X - 750}{\frac{1}{2}\sqrt{1500}}$$

is (approximately) distributed as a standard normal random variable. We may now write

$$P(X \geq 950) \;=\; P\left(\frac{X - 750}{\frac{1}{2}\sqrt{1500}} \geq \frac{950 - 750}{\frac{1}{2}\sqrt{1500}}\right)$$
$$\approx \;\; P(N \geq 10.32),$$

where N is a standard normal random variable. This last probability can be estimated numerically with a computer or a table, and this number turns out to be almost 0. The conclusion is then the following: if there is no difference in the handwriting, then the probability to guess correctly at least 950 times is about 0. Therefore, it seems safe to conclude that there is, after all, a difference in handwriting between men and women.   □

8.6 Exercises

Exercise 8.6.1. Let $X_1, X_2, \ldots$ be independent and uniformly distributed on $(0, 1)$. Let M_n be the maximum of $X_1, \ldots, X_n$. Finally, let

$$Y_n = n(1 - M_n).$$

Show, without using characteristic functions, that $Y_n \Rightarrow X$, where X has an exponential distribution with parameter 1.

Exercise 8.6.2. For $n = 1, 2, \ldots$, the random variable X_n has density f_n given by $f_n(x) = 0$ for $x < 1$ and

$$f_n(x) = \frac{n}{x^{n+1}}$$

for $x \geq 1$. Show that as $n \to \infty$, X_n converges in distribution to a random variable that takes the value 1 with probability 1.

Exercise 8.6.3. Show that the sum of n independent Poisson distributed random variables with parameter 1, has a Poisson distribution with parameter n.

Exercise 8.6.4. Use the previous exercise and the central limit theorem to show that if X has a Poisson distribution with parameter n, then

$$P(X \leq n) \to \frac{1}{2},$$

and use this to show that

$$\lim_{n \to \infty} e^{-n} \left(1 + \frac{n}{1!} + \frac{n^2}{2!} + \cdots + \frac{n^n}{n!} \right) = \frac{1}{2}.$$

Exercise 8.6.5. Let $X_1, X_2, \ldots$ be independent Poisson distributed random variables with parameter $\frac{1}{2}$. Let $\bar{X}_n = \frac{1}{n} \sum_{i=1}^{n} X_i$ be their average. Compute

$$\lim_{n \to \infty} P(\bar{X}_n \leq 1),$$

using the law of large numbers.

Exercise 8.6.6. Consider 1000 light bulbs, whose lifetime can be modelled with an exponential distribution with expectation 5 days. Approximate, using the central limit theorem, the probability that the total lifetime of the light bulbs exceeds 5200 days.

Exercise 8.6.7. Suppose that X_n has a geometric distribution with parameter $1/n$. Show that

$$\frac{X_n}{n} \Rightarrow Y,$$

as $n \to \infty$, where Y is a random variable with an exponential distribution with parameter 1.

Exercise 8.6.8. Suppose we throw a coin 10,000 times, and 5,273 heads come up. Discuss the question whether or not the coin is fair, using the central limit theorem.

Exercise 8.6.9. A bookkeeper wants to simplify his calculations by rounding off all numbers to the nearest integer number. He is interested in the sum of the first 100 rounding errors he makes. To study this, we propose the following model. We denote the rounding errors by $X_1, \ldots, X_{100}$, where the X_i's are independent and identically distibuted with a uniform distribution on $(-1/2, 1/2)$.

(a) Copute the expectation and variance of X_1.

(b) Use the central limit theorem to show that

$$P(|X_1 + X_2 + \cdots + X_{100}| > 10)$$

is approximately equal to $2P(N > \sqrt{12})$, where N has a standard normal distribution.

Exercise 8.6.10. The random variables $X_1, \ldots, X_{625}$ are independent and identically distributed with density f given by $f(x) = 3(1 - x)^2$ for $0 \le x \le 1$ and $f(x) = 0$ elsewhere. Approximate

$$P(X_1 + \cdots + X_{625} < 170)$$

in terms of the distribution function of the standard normal distribution.

Exercise 8.6.11. A random variable X is called *symmetric* if X and $-X$ have the same distribution.

(a) Give an example of a symmetric random variable.

(b) Show that a random variable is symmetric if and only if its characteristic function is real.

Chapter 9

Extending the Probabilities

In this chapter we discuss how to extend the collection of events in rather general situations. This small chapter is the bridge between probability without measure theory and probability with measure theory.

9.1 General Probability Measures

In Section 6.1 and Section 6.2 we extended the probabilities in a rather ad hoc way. It turns out that there is a more general procedure, which assigns probabilities *at once* to classes of sets that are so extensive that most of its members never actually arise in probability theory. This procedure requires some knowledge of measure theory. The purpose of the current section is to describe the ideas involved, without going into measure theory itself. This section is supposed to form the bridge between the first course in probability as set out in this book, and a future course in probability theory based on measure theory.

What does a general theory of probabilites look like? The principal set-up, with a sample space of possible outcomes does not change. This sample space can be finite, countably infinite or uncountably infinite. In the general theory, we do not restrict ourselves to $\mathbb{R}^d$ or $\mathbb{R}^\infty$; the sample space can be more exotic, for instance a space of functions with a certain property. The sample space is usually denoted by Ω, as in this book.

We have seen in this book, that it is in general impossible to assign a well defined probability to *all* subsets of Ω. We avoided this problem by restricting our attention to special classes of subsets, called events. But the route we followed, although very useful in practice and for a course on this level, is not enough for general probability theory.

In the general context, one proceeds as follows. First, one identifies a collection of subsets $\mathcal{F}$ of Ω with three basic properties:

1. $\Omega \in \mathcal{F}$;

2. $A \in \mathcal{F}$ implies $A^c \in \mathcal{F}$;

3. $A, B \in \mathcal{F}$ implies $A \cup B \in \mathcal{F}$.

♠ **Exercise 9.1.1.** Show that these properties imply that (4) $\emptyset \in \mathcal{F}$ and that (5) $A, B \in \mathcal{F}$ implies $A \cap B \in \mathcal{F}$.

Any collection $\mathcal{F}$ which satisfies (1)-(3) is called an *algebra* or a *field*. The idea behind this definition is that the collection of subsets which will receive a probability, should at least satisfy certain basic requirements: If we can speak about the probability of an event A, then it is only natural to require that also the probability of A^c be defined. And if we can speak about events A and B, then it is only natural to require that also the union $A \cup B$, corresponding to the occurrence of A *or* B is an event, and similarly for the intersection. Here are two examples.

Example 9.1.2. The collection of sets which can be written as a finite, disjoint unions of intervals in $(0, 1)$ forms an algebra.   □

Example 9.1.3. Let $\Omega = \{0, 1\}^{\mathbb{N}}$, the set of infinite sequences of 0s and 1s. A *cylinder set* is a subset of Ω of the form

$$\{\omega \in \Omega : \omega_{i_1} = k_1, \ldots, \omega_{i_m} = k_m\},$$

for $i_1, \ldots, i_m \in \mathbb{N}$ and $k_1, \ldots k_m \in \{0, 1\}$. The collection $\mathcal{F}$ defined as finite unions of cylinder sets, together with the empty set, is an algebra.   □

The next stage in the development of a general theory is to define a *probability measure* P on an algebra $\mathcal{F}$. This is a function $P : \mathcal{F} \to [0, 1]$ with the following properties:

1. $P(\emptyset) = 0$, $P(\Omega) = 1$;

2. if $A_1, A_2, \ldots$ is a disjoint sequence of sets in $\mathcal{F}$ and if $\cup_{k=1}^{\infty} A_k \in \mathcal{F}$, then

$$P\left(\bigcup_{k=1}^{\infty} A_k\right) = \sum_{k=1}^{\infty} P(A_k).$$

This last property is called *countable additivity* of P. By taking $A_{n+1} = A_{n+2} = \cdots = \emptyset$, countable additivity implies *finite additivity*:

$$P\left(\bigcup_{k=1}^{n} A_k\right) = \sum_{k=1}^{n} P(A_k).$$

The point of this part of the setup is that very often it is not too difficult to define a reasonable probability measure on an algebra. We again illustrate this with some examples.

Example 9.1.4. Consider the algebra from Example 9.1.2. If we let, for $A \in \mathcal{F}$, $P(A)$ be the sum of the lengths of the intervals that make up A, then one can show that P is a probability measure.   □

♠ **Exercise 9.1.5.** Verify this claim.

Example 9.1.6. In Example 9.1.3, we can for instance define a probability measure on the algebra by putting

$$P(\{\omega \in \Omega : \omega_{i_1} = k_1, \ldots, \omega_{i_m} = k_m\}) = 2^{-m}.$$ □

♠ **Exercise 9.1.7.** Show that this is a probability measure.

So far, the theory is nit difficult. We have defined, in an abstract setting, an algebra and a probability measure on such an algebra. Often however, an algebra does not contain 'enough' events. Therefore, we want to *extend* the probability measure to a much larger class of subsets. This is the final and most difficult step in the general theory. In order to describe this last step, we need one more definition.

Definition 9.1.8. A collection of sets $\mathcal{F}$ is called a *σ-algebra* if it is an algebra, and if it is also closed under the formation of *countable* unions:

$$A_1, A_2, \ldots \in \mathcal{F} \text{ implies } \bigcup_{k=1}^{\infty} A_k \in \mathcal{F}.$$

A probability measure on a σ-algebra is defined as on an algebra, the only difference being that in (2), we do not need to require that the countable union is in $\mathcal{F}$, as this is automatically the case in a σ-algebra.

Given an algebra $\mathcal{F}$, we define the σ-algebra *generated* by $\mathcal{F}$, as the smallest σ-algebra that contains $\mathcal{F}$. This generated σ-algebra contains all sets that can be obtained from sets in $\mathcal{F}$ by taking countably many unions, intersections and complements.

♠ **Exercise 9.1.9.** Can you prove this last statement?

The last step of the general theory may now be compactly stated in the following way:

Theorem 9.1.10 (Extension theorem). *A probability measure on an algebra has a unique extension to the generated σ-algebra.*

This theorem contains two statements: we can extend the probability measure to a much larger (in general, at least) class, and we can do this in only one way, there is uniqueness. The proof of this theorem is quite lengthy and difficult, and it would not be appropriate to include this in this book. Note however that this construction immediately assigns a well defined probability to all sets that can be expressed as countable unions, intersections and complements of sets in the algebra one starts with.

So, for instance, if we want to make a satisfactory model for infinitely many coin flips, then we start with the algebra from Example 9.1.3, and define the 'obvious' probability measure on this algebra. The extension theorem then tells us that this *uniquely* defines a probability measure on the σ-algebra generated by this algebra, and this σ-algebra contains all sets that could ever be of interest to us. If, on the other hand, we want a satisfactory model for choosing a random point in $(0, 1)$, then the algebra in Example 9.1.2 is appropriate as a starting point.

Hopefully, this short description provides enough motivation to learn more about measure theory and probability theory based on this.

Appendix A

Interpreting Probabilities

In this appendix, we would like to pay some attention to the practical use and interpretation of the theory that we have discussed in this book. Probability theory is one of the most frequently used branches of mathematics, and perhaps also one of the most *abused*. This is not the place to discuss at length the philosophical issues which arise from using an abstract theory in practice, but it is worth, I think, to point out a number of things.

The very first thing to realise, is that what we have done in this book, is to set up a mathematical *model*, which hopefully gives a reasonable description of some aspects of daily life. For instance, we all know that when we throw a coin repeatedly, that the relative frequency of heads converges to $\frac{1}{2}$. In that light, it is reassuring that in our model we can prove laws of large numbers that agree with this empirical fact. It is, of course, not the case that we have *proved* anything about daily life. An empirical law is not part of mathematics, and there is nothing to prove there. All we can conclude from the fact that experience agrees with theory, is that the model is a good model, as far as this particular aspect is concerned.

When we set up a model, the role of this model is to *describe* a certain aspect of reality. If it turns out that the model behaves satisfactorily (as is clearly the case with repeated coin flips), then we might even turn to our model to make *predictions*. This is for instance what casino's do. Casino's know that the laws of large numbers work very well in real life, so by making their games slightly unfavourable for their customers, casino's are bound to make a lot of profit. Similarly, airline companies purposely overbook their flights by a certain percentage. They simply know, according to the law of large numbers again, that they will have to reimburse a certain number of people because of this, but overbooking makes sure that the aircrafts will be full. A computation then easily shows which overbookpercentage is optimal. Insurance companies also use the law of large numbers to calculate risks.

In all these cases (you can easily come up with other examples, we guess) it turns out that probability theory is well capable of describing and predicting. Why is this so? Why is probability theory so incredibly useful?

When we construct a mathematical model involving randomness, this is very often done because we do not have full information about the process to be modelled, and we interpret the unknowns as being random. In other words, the things that we can not be sure of, are modelled as being random. The number of customers in a shop during an hour is something we cannot foresee, and therefore we model this by a random variable. It has been a long term philosophical question whether or not this number is truly random, or only perceived as being random by us, humans. I do not think that this is a very meaningful question, for the simple reason that I would have no idea how to define the word 'random' outside a mathematical context. Whenever we think of processes as being random, like winning a lottery, throwing a 6 with a die, or meeting someone on the street, as soon as we associate this apparent randomness with probabilities, we are, in fact, already working and thinking in a mathematical model. For instance, when you claim that the probability of a certain rare event is no more than 1 out of 10,000, than in fact you talk about a *model* which is supposed to describe this event. This can easily lead to misunderstanding and abuse of probability theory. I do not think it is appropriate to dismiss very remarkable events by saying that probability theory tells you that rare events will occur eventually, no matter how small their probability. Saying this is turning things upside down: When you want to say something about such a very rare and remarkable event, you first make a model, in which this event has nonzero, but small probability. If the event then occurs, you might be tempted to appeal to the model and say: 'well, it is not so remarkable that the event occurs, because my model assigns positive probability to it'. But this is a circular reasoning, since you have designed the model yourself, including the nonzero probability. Hence I would rather not dive into troubled water, and prefer not to make any statement about randomness in real life, whatever that may be, and no matter how one should or would try to define it.

This means that I do not think that the role of probability theory is to model random events. As I have tried to explain, this would be a meaningless statement, given the fact that I do not know what the word 'random' means. Instead, I prefer to see probability theory as a way to deal with complex phenomena which would have been hard or even impossible to describe otherwise. The use of randomness in a mathematical model is a *conscience choice* of the person who designs the model, and has, in principle, nothing to do with the driving forces that are behind the process described by the model.

I think that this is a very important point. One can ask when this is appropriate, I mean when it is appropriate to use randomness in modelling a phenomenon. I think there is no general, clear-cut answer to this question, but I have a rather pragmatic view: I consider randomness in modelling appropriate when experience with the model shows that the model works. Our experience with modelling with randomness shows that it works very often and very well. But this point of view implies that I am quite critical towards a fair number of examples where probability theory is used to draw conclusions. For instance, when someone claims that the probability of life elsewhere in the universe is large, because there are so many planets on which life could possible emerge, then I do not think that this is a statement that can be taken seriously. The reason is simple: there is no way to check whether or not the model used is, in fact, describing things well. We cannot check whether the model makes sense. I have similar problems with probabilistic statements about the possibility of a big flood in the coming 100 years or so. How can I convince myself that the model used here is any good?

So stay alert when people make probabilistic statements about future developments. Probability theory is a very useful tool, but should be used with care and attention.

Appendix B

Further Reading

After the introduction in this book, there are a number of ways to proceed. If you are interested in the foundations of probability theory in terms of measure theory, then Billingsley's excellent *Probability and measure* (3rd edition Wiley 1995) is a very good idea. In this monumental book, all necessary measure theory is developed along the way, motivated by probabilistic questions. Another good possibility in this direction is *Foundations of modern probability* by Kallenberg (Springer 2002). Both these choice are quite demanding.

Breiman's classic *Probability* (Addison-Wesley 1968) is still a very good choice to continue reading, and a little easier to digest. For a very original approach you can try *Probability with martingales* by Williams (Cambridge University Press 1992) which also includes the necessary details of measure theory. For a book of roughly the level of this book, consult *Probability and random processes* by Grimmett and Stirzaker (Oxford 1993). Feller's classics *An introduction to probability theory and its applications I and II* (Wiley 1978) are still very much worth the effort. More recent is *A user's guide to measure theoretical probability* of Pollard (Cambridge University Press 2001), also aiming at an audience without measure-theoretical background, but making a different choice by providing measure-theoretical details (without proofs) along the way. Durrett's *Probability: theory and examples* (Duxbury Press 1995) is a somewhat demanding but rewarding book, containing a lot of interesting examples illustrating the theory. Perhaps you will find Gut's recent *An intermediate course in probability* (Springer 1995) useful as a follow-up on the current book.

For a very pleasant introduction to some simple but beautiful stochastic processes, we recommend *Markov chains* by Norris (Cambridge University Press 1997). For generalisations of the Poisson process in the framework of so-called renewal processes, it is a good idea to read Karlin and Taylor *A first course in stochastic processes* (Academic Press New York 1975). There even is a very nice little book solely devoted to Poisson processes in arbitrary spaces, Kingman's *Poisson processes* (Oxford 1993). For more details on branching processes, a good option is *Branching processes* by Athreya and Ney (Springer 1972). Many of the above-mentioned books contain a large number of references to the literature.

Appendix C

Answers to Selected Exercises

Chapter 1: 1.7.3 (a) $1/6$, (b) $1/9$, (c) $1/12$; 1.7.4 $4\binom{48}{9}/\binom{52}{13}$; 1.7.5 (a) $5/12$, (b) $5/12$, (c) $4/11$; 1.7.6 (a) $29/32$, (b) $9/16$, (c) $81/256$; 1.7.8 $17/27$; 1.7.9 $\binom{13}{5}\binom{13}{8}\binom{52}{13}^{-1}$; 1.7.11 $1/n$; 1.7.12 $333/1000$, $1/5$, $9/1000$; 1.7.14 $1/16$; 1.7.17 no; 1.7.19 $(8/21)/$ $(8/21 + 3/28)$; 1.7.21 $\binom{5}{2}\binom{15}{2}\binom{20}{4}^{-1}$; 1.7.26 $1/3$; 1.7.33 $1/3$, $1/2$; 1.7.35 (a) $1/3$, (b) $1/5$.

Chapter 2: 2.7.3 $1 - \frac{17}{2}e^{-3}$; 2.7.9 (b) $P(X = k|D) = 1/3$ for $k = 6, 10, 12$; 2.7.14 (a) e^{-1}, (b) $1/2 + 1/2e^{-2}$, (c) 1; 2.7.17 r/p; 2.7.18 (a) $18/38 + 20/38(18/38)^2 = 0.5917$; 2.7.20 (a) $3/5$, (b) $2/3$, (c) $E(X|N = n) = n/2$, $E(X) = 1/3$; 2.7.21 (c) no, (d) $6,5$; 2.7.22 $E(X|Y = 4) = 2.5$, $E(Y|X = 2) = 3$; 2.7.28 (b) $7/30$; 2.7.30 the marginals are bonimial with parameters n and p_i; 2.7.33 $k/2$; 2.7.41 (a) $Z = 2X - 5$, (b) $E(X) = -5/3$.

Chapter 4: 4.3.1 (c) the random variables are not independent.

Chapter 5: 5.12.2 (a) $t = \sqrt{2}$, (b) $4\sqrt{2}/5$, (c) $3/4$; 5.12.3 a and b satisfy $a + \frac{1}{3}b = 1$ and $\frac{1}{2}a + \frac{1}{4}b = \frac{3}{5}$; 5.12.4 (a) 3, (b) $3/4$, (c) $7/8$, (d) $3/80$; 5.12.5 (a) $1/2$, (b) $\pi/2$; 5.12.7 (a) 60, (b) $11/32$; 5.12.12 (a) $\lambda/(\lambda + \mu)$; 5.12.18 (a) $f_Y(y) = ye^{-y}$ for $y > 0$, (c) $E(X|Y = y) = y/2$, $E(X) = 1$; 5.12.19 (a) $f_X(x) = e^{-x}$ for $x > 0$, $f_Y(y) = (y + 1)^{-2}$ for $y > 0$, (c) $2/(y + 1)$; 5.12.20 yes; 5.12.21 (c) $1/x$; 5.12.22 (a) $f(x, y) = 2(x^2 - 4y)^{-1/2}$ for $X \in (0, 2)$. $y \in (0, 1)$, $x^2 > 4y$; 5.12.23 $f_Y(y) = 2\pi^{-1}(1 - y^2)^{-1/2}$ for $y \in (0, 1)$; 5.12.25 $f(w, v) = -\frac{1}{2}\log w \cdot v^{-1/2}$ for $w, v \in (0, 1)$; 5.12.31 $1/8$; 5.12.33 $(2\lambda)^{-1}$; 5.12.34 $f(z) = z$ for $0 < z < 1$ and $f(z) = 2 - z$ for $1 < z < 2$; 5.12.36 $f_Z(z) = e^{-z}$ for $z > 0$.

Chapter 7: 7.5.1 (a) $P(X \leq 10)$ where X has a Poisson distribution with parameter 9, (b) e^{-1}, (c) $e^{-4\frac{1}{2}}$; 7.5.2 $P(N(s) = k|N(t) = n) = (\frac{s}{t})^k(1 - \frac{s}{t})^{n-k}\binom{n}{k}$, the binomial distribution with parameters n and s/t.

Chapter 8: 8.6.5 1.

Index

σ-algebra, 191
    generated, 191

Abel's theorem, 65
algebra, 190
arc-sine law, 80

ballot theorem, 77
Banach-Tarskii paradox, 97
Bayes' rule, 17, 26
binary expansion, 96, 144
binomial coefficient, 6
branching process, 153
    extinction of, 155, 156
bridge, 9
Buffon's needle, 141

Cantor set, 157
Cauchy-Schwarz inequality, 52
central limit theorem, 90, 185
    local, 93
characteristic function, 176
    and independence, 177
    expansion of, 181
    of binomial distribution, 178
    of Cauchy distribution, 184
    of exponential distribution, 178
    of normal distribution, 182
    of Poisson distribution, 178
Chebyshev inequality, 52, 137
coin tossing, 10
    fair, 10
    general, 21, 43
combinatorics, 6
complement, 4

conditional, 59, 133
    density, 134
    distribution, 59
    distribution function, 133
    expectation, 59, 134
    probability, 13, 59
continuity theorem, 179
contour integral, 184
convergence, 80
    in distribution, 80, 174
    weak, 173
countable, 2
countably infinite, 2
counting, 6
covariance, 51, 128
    of bivariate normal distribution, 128

darts, 100, 101
de Méré's paradox, 34
density, 102
disjoint, 5
    pairwise, 5
distribution, 42, 108
    binomial, 39
    bivariate normal, 118, 135, 140
    Cauchy, 110
    double-exponential, 139
    exponential, 110
    gamma, 124, 164
    geometric, 39
    multinomial, 71
    negative binomial, 40
    normal, 109
    Poisson, 39

truncated exponential, 169
uniform, 108
distribution function, 38, 107
joint, 55, 116
properties of, 40
DNA profile, 25
dominated convergence, 81, 104
drawing, 7
with replacement, 7
without replacement, 7

envelope problem, 63, 114
estimate, 171
event, 3, 102
expectation, 46, 65, 111, 125
of a function, 47, 126
of a mixture, 129
of a sum, 48, 125
of binomial distribution, 53
of exponential distribution, 112
of Gauchy distribution, 114
of geometric distribution, 53
of negative binomial distribution,
69
of normal distribution, 113
of Poisson distribution, 46
of uniform distribution, 113
experiment, 3, 102
exponential series, 24
extension theorem, 191

generating function, 64, 154
of geometric distribution, 64

independence, 19, 43, 56, 66, 118, 145,
160
pairwise, 20
indicator function, 58
infinite repetitions, 143
inter-arrival time, 163
intersection, 5
inversion theorem, 178
island problem, 25

Jacobian, 121

lack of memory, 69, 139
law of large numbers, 28, 137, 183
strong, 148–150, 170
weak, 85, 87
left to right crossing, 23

marginal distribution, 55, 116
Markov inequality, 52, 151
master mind, 7
mixture, 129
model, 159

networks, 22
dual, 23
numerical integration, 137

partition, 16
Poisson process, 164
in higher dimensions, 172
inhomogeneous, 172
thinning of, 172
predictions, 193
probability density, 102
probability mass function, 3, 38
joint, 55
probability measure, 3, 102, 190
countable additivity of, 11, 190
finite additivity of, 12

random variable, 37, 106, 147
continuous, 106
discrete, 106
functions of, 119
symmetric, 188
random vector, 55, 115
continuous, 116
random walk, 75, 153
recurrence of, 153
reflection principle, 77
regular, 122
relative frequency, 1
roulette, 89

sample space, 3, 102
secretary's problem, 72

set, 6
    finite-dimensional, 146
    ordered, 6
    small, 146, 147
    unordered, 6
Simpson's paradox, 32
size bias, 27
St. Petersburg paradox, 46
standard deviation, 50
stationarity, 159
statistical application, 171, 186
Stirling's formula, 80
sum, 58
    of Poisson distributions, 58
    of two random variables, 58, 123

uncountable, 96
union, 5

variance, 50, 65, 127
    of a sum, 50, 128
    of binomial distribution, 53
    of normal distribution, 127
    of Poisson distribution, 53
    of uniform distribution, 127

waiting time, 103, 163
waiting time paradox, 168
well defined, 45